The AI Economy

The AI Economy

Work, Wealth and Welfare in the Robot Age

ROGER BOOTLE

nb

First published in Great Britain in 2019 by Nicholas Brealey, an imprint of John Murray Press. An Hachette UK company.

British Library Cataloguing in Publication Data: a catalogue record for this title is available from the British Library.
Library of Congress Catalog Card Number: on file.

Hardback ISBN: 978 1 473 69616 7
Trade paperback ISBN: 978 1 473 69615 0
eBook ISBN: 978 1 473 69617 4

Typeset by Cenvo® Publisher Services

Printed and bound in Great Britain by Clays Ltd, Elcograf S.p.A.

John Murray Press policy is to use papers that are natural, renewable and recyclable products and made from wood grown in sustainable forests. The logging and manufacturing processes are expected to conform to the environmental regulations of the country of origin.

Nicholas Brealey Publishing
John Murray Press
Carmelite House
50 Victoria Embankment
London EC4Y 0DZ, UK
Tel: 020 3122 6000

Nicholas Brealey Publishing
Hachette Book Group
Market Place Center
53 State Street
Boston, MA 02109, USA
Tel: (617) 263 1834

www.nicholasbrealey.com

Contents

Preface vii

Prologue: The Robot Age 1

Part I: Man and machines: past, present, and future 13

1 The ascent of man 15

2 Could this time be different? 41

3 Employment, growth, and inflation 69

Part II: Jobs, leisure, and incomes 95

4 Work, rest, and play 97

5 The jobs of the future 127

6 Winners and losers 161

Part III: What is to be done? 193

7 Encourage it, or tax and regulate it? 195

8 How the young should be educated 219

9 Ensuring prosperity for all 245

Conclusion 279

Epilogue: The Singularity and beyond 285

Bibliography 295

Notes 301

Index 329

Preface

The genesis of this book was straightforward. Over the last three years I have been overwhelmed by accounts of the impending shock to the economy and society from the spread of robots and the advances of artificial intelligence (AI). And I have met umpteen people – and businesses – who are seriously concerned about this, in some cases concerned to the point of panic. The ranks of the concerned apparently even include many people in leading AI firms who are worried about the possible impact of AI on humanity but daren't voice their fears in public because of the possible implications for their careers.[1]

It is evident to me that this is one of the greatest economic issues of our age. Moreover, it promises to be one of the greatest social issues as well. Indeed, it threatens to overwhelm everything else that we are concerned about. Jim Al-Khalili, Professor of Physics at Surrey University and President of the British Science Association, has recently said that AI is more important than all other big issues facing humanity, "including climate change, world poverty, terrorism, pandemic threats and anti-microbial resistance."[2] Whether or not he is correct about the ranking compared to, say, climate change and terrorism, this makes robots and AI seem wholly bad. In fact, plenty of people (including, interestingly, Professor Al-Khalili) think that they have extraordinary capacity to bring benefits to humanity.

So, whether the implications are dastardly or beneficial, I have wanted to understand what is happening with robots and AI and to consider what the consequences might be. This book is the result of my investigations.

After spending more than a year immersed in the literature, I now know my way around the subject reasonably well. Yet readers should not be concerned that my new-found knowledge has turned me into a geek. Rest assured that it has not detached me from my bearings.

I approached this subject without the advantage of great technical knowledge. In fact, I should confess that the truth is worse than this. Before my immersion in the AI literature, my children regarded me as something of a technophobe. And I am sure that the staff at Capital Economics (the company I founded) would readily have concurred with this judgment. My only edge has been a knowledge of economics and a lifetime spent thinking and writing about economic issues.

Not that this brings any direct advantage in understanding the technical matters at the bottom of this subject. And indeed, it did not prevent me from being utterly flummoxed and bamboozled by much of what the technical experts have to say. I may now claim to have reached a reasonable level of understanding, but I have spent many a long hour swathed in the proverbial hot towels, struggling to comprehend what the robotics and AI experts were saying.

Could my very absence of prior technical knowledge and understanding bring some sort of *indirect* advantage to set against the obvious disadvantages? Perhaps. At least it put me in the same starting position as most of my readers who, just as I once was, feel battered and bemused by the subject.

Waves of technobabble seem to flood over us whenever the words "robots" and "artificial intelligence" appear. To read about these subjects is to immerse yourself in a sea of waffle, wonder, and worship at the altar of "technology." You risk drowning in the onrush of loose language, flabby concepts, crude extrapolation, impenetrable jargon and lack of perspective, all wrapped up in an aura of supposed inevitability.

And yet something truly amazing *is* happening in the world of technology, not just increasing digitization, or the development of nanotechnology, biotechnology and 3D printing, but also with regard to robots and AI. In this cocktail of ingredients for a technological revolution it is AI that stands out. It may offer the greatest benefits, but it also seems the most threatening to individual human beings and to society as a whole. For it seems to penetrate deep into the human realm and to pose fundamental questions about who we are and what we may become.

My task here is not only to survive the deluge of technobabble myself and to ensure that my readers do not drown in it but, more importantly, to salvage the many nuggets of truth amid this sea of exaggeration and to draw out the possible implications for our future.

These possible implications are wide-ranging. The experts on AI do not confine themselves to the merely technical. Indeed, they range far and wide across the whole terrain of economics, social structure, politics, and even the meaning of life. In the process, they produce conclusions that are both baffling and potentially terrifying for individuals, companies, and governments.

These three groups find themselves at a loss to know what to think about the momentous issues at stake – let alone what to do in the face of them. It is here that the background, training, and experience of an economist can profitably be brought to bear. At least, I hope they can. In the end, it is you, the reader, who must be the judge of this.

I should emphasize, though, that you won't find here much enlightenment about the technical details or the essential nature of robotics and AI. Readers who want these things will have to look elsewhere. This is a book about the economic consequences of robots and AI. It is my attempt to bring clarity about these consequences to anyone who is potentially affected by them – clarity about how to regard these developments, and clarity over the choices that face them about what to do. By the way, these issues are so momentous that in this instance "anyone" really amounts to "everyone."

Yet these issues are so complex and intertwined with all sorts of things whose future cannot be known that, however much clarity can be brought to the subject, there can be no certainty. When peering into the future we simply have to skin our eyes and try to make out, as best we can, what shapes lie before us.

As with my previous books, I don't think that this profound uncertainty is an excuse for saying and doing nothing. After all, everyone has to make decisions that rest to a considerable extent on a view of the future. This applies to individuals as much as to businesses and governments. We cannot escape from uncertainty and we cannot put off all decisions until everything is clear. We simply have to do the best we can.

Inevitably, in the course of researching and writing the book I have accumulated many debts of gratitude. Professor Robert Aliber,

Dr Anthony Courakis, Julian Jessop, Gavin Morris, George de Nemeskeri-Kiss, Dr Denis O'Brien, Dr Alya Samokhvalova, Christopher Smallwood, Martin Webber, and Professor Geoffrey Wood kindly read various versions of the text and gave me their comments. I am most grateful to them all, and also to the participants at a round-table discussion in Vienna in December 2018, organized and hosted by the Austrian bank OeKB.

I was fortunate to enjoy the services of a research assistant, Moneli Hall-Harris, who made my task so much easier. Many staff members at Capital Economics helpfully provided data and charts, and others supplied critical comments, especially Andrew Kenningham, Mark Pragnell, Vicky Redwood, Nikita Shah and Neil Shearing. Moreover, I am grateful to Capital Economics for permission to include here the results of some of the research studies it has published over recent years, particularly on the subjects covered in Chapter 1.

My PA, Holly Jackson, was invaluable both in managing the typescript and keeping me on the straight and narrow with regard to my activities at Capital Economics and everything else that I have been involved with.

Last but not least, I owe a debt to my family for putting up – yet again – with my distraction and absorption in writing a book.

None of the above, nor anybody else mentioned in the following pages, is at all responsible for any errors of commission or omission. These remain the responsibility of the author alone.

Roger Bootle
London, March 2019

Prologue: The Robot Age

"It looks less like *Battlestar Galactica* and more like the Fall of Rome."

David Gunkel[1]

"The saddest aspect of life right now is that science gathers knowledge faster than society gathers wisdom."

Isaac Asimov[2]

Bubbling enthusiasts for robots and artificial intelligence (AI) gush technical about how a new revolution is going to transform our lives. But it usually isn't clear whether the transformation is going to be favorable or unfavorable. What I read from the geeks is a mixture of two distinctly different visions: first, the idea that we all face a terrible future – involving poverty, loss of self-worth, or even annihilation – as our creation, the robots, take over; second, the idea that the revolution is going to enrich us all and free humanity from drudgery.

In the process of discussing the implications of robots and AI, many writers of a technical background have ranged far and wide into the territory of macroeconomics and public policy. Take this, for example, from the AI visionary Calum Chace:

> *... as the machines will be more efficient than the humans they replaced, and increasingly so, as they continue to improve at an exponential rate. But as more and more people become unemployed, the consequent fall in demand will overtake the price reductions enabled by greater efficiency. Economic contraction is pretty much inevitable, and it will get so serious that something will have to be done.*[3]

Similar views have been expressed by leading tech entrepreneurs, including Bill Gates, the founder of Microsoft, and distinguished scientists such as the late Sir Stephen Hawking, the discoverer of black holes and much else besides.[4]

Yet some cynics, admittedly mostly from outside the ranks of the AI experts, seem to think that this is all overhyped and that in essence the economic and social changes that we face as a result of the spread of robots and AI will either be nugatory, or a continuation of the sort of thing that we have experienced pretty much continuously since the Industrial Revolution and, as such, will be profoundly beneficial to humanity. Some see the specters unleashed by fetid speculation about the implications of AI as reminiscent of the gigantic fuss about the Y2K computer bug which, in the end, came to naught.

Five visions of the future

So, to put it mildly, this is a far from settled issue. The conflicting views about our fate in a robot-and AI-dominated future can be pithily summarized as follows:

- Nothing different.
- Radically bad.
- Radically good.
- Catastrophic.
- The key to eternal life.

Charting a path through these different possible futures is the main purpose of this book. I cannot, and should not, try to steal my own thunder just as the argument is about to begin. So, I will leave discussion of the first three possible outcomes to succeeding chapters. But I must briefly say something about the fourth and fifth possibilities now.

To anyone unfamiliar with the literature on this subject the words I have chosen to describe the fourth and fifth visions of the future – namely catastrophic and the key to eternal life – will seem hyperbolic. But anyone who has delved into the literature will recognize them as anything but. The techies argue that once human-level AI is achieved, superhuman AI will be almost inevitable. A digital brain can be copied without limit and, unlike the human brain, it can be speeded up.

This then leads on to the idea that the AI revolution potentially amounts to the last human advance. Once we have created artificial intelligences greater than any human one, they will then create still greater intelligences, completely beyond our ken and outside our control. And so on, and so forth. To these new forms of intelligence, we will be not only inferior but also worthless, if not actually an encumbrance. They could readily decide to destroy us. The late Sir Stephen Hawking told the BBC in 2014: "The development of full artificial intelligence could spell the end of the human race."[5]

Similarly, the distinguished Cambridge scientist Martin, Lord Rees, the Astronomer Royal, has described the point at which AI achieves superintelligence as "our final hour." He sees the period when human intelligence dominated the world as a blip.[6]

The time at which some form of AI becomes more intelligent than humans is widely referred to in the literature as "the Singularity." As and when it happens, the consequences are likely to extend well beyond individuals. Murray Shanahan, the Professor of Cognitive Robotics at Imperial College, London, encapsulated the views of many AI experts when he wrote:

> *"By analogy, a singularity in human history would occur if exponential technological progress brought about such dramatic change that human affairs as we understand them today came to an end. The institutions we take for granted – the economy, the government, the law, the state – these would not survive in their present form. The most basic human values – the sanctity of life, the pursuit of happiness the freedom to choose – these would be superseded."*[7]

But the world of the Singularity is far from being entirely negative for humanity. Indeed, for Ray Kurzweil, the high priest of AI enthusiasts, it is just the opposite. He sees a fusion between humans and AI that effectively enables us to "upload" ourselves into a nonmaterial form and thereby secure eternal life.[8] (I suppose this *is* a more positive vision for us all, isn't it? I don't know about you but, personally, being "uploaded" into some form of AI for eternity does not exactly appeal to me.)

To someone of my ilk, even without the "uploading" and the prospect of eternal life, to read about the capability of AI and the fate of humanity after the Singularity is to plunge into a world that seems like science fiction. Nevertheless, as I shall show in the Epilogue, I do not dismiss such ideas. How could I? When some of the greatest scientific minds of our age, such as Sir Stephen Hawking and Lord Rees, have taken these prospects seriously, I am hardly in a position to disparage them.

But I am profoundly conscious of a dislocation between the apparently science fiction world of the Singularity and the day-to-day advances of robotics and AI that have here-and-now effects on the economy. These require decisions from both companies and individuals in pursuit of their own interests, and governments in pursuit of the public good.

What influence should the Singularity have on these decisions? Among others, John Brockman, a "cultural impresario" who has connections with most of the world's leading scientists and thinkers about AI, thinks that current decisions by all these actors should be strongly influenced by the coming revolutionary change. He writes: "One doesn't need to be a super intelligent AI to realize that running unprepared toward the biggest event in human history would be just plain stupid."[9]

And Stuart Russell, Professor of Computer Science at the University of California, has alleged that to fail to prepare for the Singularity would be complacent or even downright irresponsible. He has written: "if we received a radio signal from a more advanced alien civilization saying they'd arrive here in sixty years, you wouldn't shrug and say, 'Eh, its sixty years off.' Especially not if you had children."[10]

I strongly disagree with this view. I would not immediately rush to make major decisions on the basis of such a "radio signal." First, I would want to know that the message really was from an alien civilization and I would wonder what the source's record was for correctly preannouncing its actions or forecasting events. If this was a first "message," of course, there would be no such record. That would put me on my guard. And I would be acutely aware that the famous Orson Welles radio broadcast of H. G. Wells's futuristic novel *The War of the Worlds*, aired on October 30, 1938, claiming that aliens from Mars had invaded New Jersey, terrified thousands of Americans and caused widespread panic.

Then, if I thought the message was to be believed (Orson Welles and H. G. Wells notwithstanding), I would wonder about what we could do. Should we prepare for war or plan a welcoming party? And suppose that the message said that they were coming in 200 years, or even 500, or just some time maybe? What we should do could be radically different depending upon how imminent or distant their arrival was.

In reality, those who claim that the Singularity is near are not beings from some more advanced, alien, civilization but rather the ultra-enthusiasts of a cult here on earth. And, for all their enthusiasm, although they may turn out to be right, there are good arguments (which I will discuss at the end of the book) as to why they will prove to be wrong.

And timescale really matters. Lord Rees, whom I quoted earlier as believing in the burgeoning and irresistible power of AI, has suggested that machines will *probably* have taken over "within a few centuries." Never mind impossibility, if the world of the Singularity is as long delayed as Lord Rees believes that it probably will be, then it is possible that the human race may be done for much earlier, by nuclear war, asteroid collision, pandemic, or heaven knows what. Or perhaps before then it will find salvation some way or other.

On a cosmic scale, a few hundred years may be a mere heartbeat but for the designers of public policy, never mind for all of us as individuals, it might as well be infinitely distant. A focus on the Singularity would lead decision-makers up the garden path with regard to day-to-day issues in the here and now. Moreover, these day-to-day issues are likely to be with us for many years to come. Indeed, day-to-day may readily slide into decade-to-decade, and even century-to-century. Fashioning our lives and shaping public policy now to prepare for the Singularity at some unspecified point in the future would be an expensive folly. Worse than that, it would obscure the palpable advances of robots and AI in the near future, causing us to be dangerously unprepared for what lies immediately in front of us.

So, right at the beginning of the book, I have taken a major decision. The world of the Singularity and all its implications I have consigned to the end. (You might well think that this is precisely where both the Apocalypse and the promise of eternal life necessarily belong.) Everything else in the book refers to the anterior world – the immediate world – in which although robots and AI become more important, the human race is not overtaken by AI, let alone wiped out by it. Nor are we uploaded into cyberspace.

But this does not mean that we should downplay – let alone ignore – what robots and AI are going to do to the economy and society. The changes they will bring will be profound. We may or may not be heading for the Singularity, but we are heading into the AI economy. This book is about what this will be like for humanity.

Terms and definitions

As with all analytical issues where things are developing rapidly, there are some tricky questions about terms and definitions. What do we mean by "robots" and "artificial intelligence"? The word "robot" is believed to have been first coined in the 1920 play *R.U.R.* (standing for *Rossum's Universal Robots*) by Czech science fiction writer Karel Čapek. Its linguistic roots seem to lie with the word *robota*, meaning obligatory work, and *robotrick*, meaning "serve."[11]

Whatever its origins, the word "robot" has entered not just the language but also our imaginations. It naturally conjures up for us a metallic figure shaped roughly like a human, with a head, two hands, arms, and legs. Yet many things that we might want to call "robots" are not shaped at all like this. We should think of robots as mechanical devices that can be programed to act in certain ways, without necessarily looking, or trying to behave, like humans. I will use the word "robot" to mean all such devices, whatever their shape and appearance.

The term "artificial intelligence" was coined in 1955 by John McCarthy, a professor of mathematics at Dartmouth College in the USA. Along with colleagues from MIT, Bell Laboratories, and IBM, he set out to "find out how to make machines use language, form abstractions and concepts, solve kinds of problems now reserved for humans, and improve themselves."[12]

John Brockman has suggested that the term "artificial intelligence" is unhelpful. He prefers "designed intelligence." Whether or not he is right about the merits of this suggestion, the term "artificial intelligence," or AI, is so well established in the literature and in public discussion that it would be confusing and unhelpful to change it now. Accordingly, I am sticking with it.

This may settle the nomenclature, but it still leaves us with tricky issues of definition. Actually, the borders between ordinary machines and robots, and between robots and AI, are unclear. Is a washing machine a robot? We would not normally want to refer to it in that way. But is this because washing machines aren't shaped like humans and don't move about? Equally, by programing what we would ordinarily accept to be a robot to do certain things that a human could do aren't we imbuing it with a degree of intelligence – indeed "artificial intelligence"?

In fact, there is a large literature on this subject, discussing what constitutes a robot and what qualifies as AI, as well as the links between the two. I intend not to burden the reader with a definitional diatribe here. I invite those who are interested to delve into the literature.[3] Throughout the book I frequently use the term "robots and AI" as a shorthand way of referring to the whole genus. I suspect that readers will readily understand what I mean without having to reach for an AI lexicon or agonizing about definitional boundaries.

Purpose

Although I hope that economists will find in these pages much to interest and engage them, the book is not written primarily for them but rather for the intelligent general reader. For some people, contemplating the issues raised by the spread of robots and AI will be merely a matter of curiosity and interest. I hope to provide them with food for thought. But for many readers the subject matter of this book will go right to the heart of some of their major anxieties about the future and address key matters on which they will need to make decisions. This includes many people working in robotics and AI and sectors closely affected by them. The book's purpose is to aid understanding and hence to improve decisions, while also enhancing confidence about the future.

For individuals, the key issues will revolve around work, but there are also important issues concerning leisure time, retirement, and parental responsibilities:

- Will the sort of job that they currently have, or plan to have, enjoy a bright future or is it destined for extinction, or something in between?
- What sort of skills should they seek to acquire and develop in order to maximize their employability and earning capacity in the future?
- Should they anticipate long spells, voluntarily or involuntarily, without work?
- Even when they are in employment, will hours of work be noticeably shorter?
- Should they prepare for a much longer retirement?
- As parents, how should they seek to educate their children to prepare them for both work and leisure in the AI economy?

For businesspeople the primary questions are related to these same issues, but they have a slightly different focus:

- Which business activities have a bright future in the new world, which could be wiped out by the effects of robots and AI, and which could have a future somewhere in between?
- In which areas should businesses be investing heavily in robots and AI?
- In which activities should they be investing in skilling up their workforce, and, if so, how?
- In which activities should they plan to replace workers with robots and AI? In which should they regard the need for labor as pretty much impervious to robots and AI? And in which should they expect humans to work closely together with robots and AI?
- What sort of new industries and activities may spring up?

For governments, too, and all those interested in public policy, these same questions are crucial, but their key concerns also have a different gloss and emphasis:

- How will the performance of the economy respond to robots and AI and what challenges will they pose for economic policy?
- Should governments seek to encourage or restrict the spread of robots and AI? And, if so, how?
- How do the law and regulatory systems need to be changed to accommodate robots and AI?
- What is the role of the state in reforming the education system in order to make it fit for the new world?
- Should governments be preparing a radical revamp of the tax and benefit system in order to offset a possible widening in inequality? And, if so, what measures should they be prepared to take?

Shape and structure

Because the concerns of these three groups are closely related, rather than separate out chapters for individuals, businesses and policy-makers, I separate them by subject, so that each chapter should be of interest to all three groups.

The structure of the book is simple, but it still needs a word of explanation. Although I examine the issues here in neat, self-contained chapters, they are all closely related. Moreover, the interrelationships between the subjects discussed go in different directions simultaneously. So, there is a real problem of where to break into the circle and how to structure the analysis.

Researching the book and then writing it has been a voyage of discovery. I hope that you will experience something similar as you read it. But the reader's journey should be both shorter and more structured than the author's. When the latter begins his journey and wanders hither and thither in search of heaven knows what, he does not know where his enquiry will end. But once the journey is over, he knows both the destination that he wishes to bring the reader to and the most direct route by which to reach it.

Consequently, it makes sense that readers are presented with a structured approach to the subject. Yet this means that they are bound to have many "but what about so-and-so" moments as they read about one aspect of the subject, conscious of the interconnections with other aspects that haven't yet been dealt with and not sure that they will ever be. As far as possible, I try to help readers to maintain their patience and enhance their understanding by indicating where issues that have been apparently ignored or glossed over are dealt with later in the book.

Whether ordinary individuals, businesspeople, or workers in government and public policy, many readers will doubtless be itching to get straight to the nitty-gritty questions concerning the effects of robots and AI on the various aspects of their lives and activities described above. But they will need to hold their horses for a while. Any attempt to speculate about the future of work, incomes, education, leisure, and a whole host of other things that may be affected by robots and AI will be meaningless without an understanding of the macro environment. Indeed, it is precisely the lack of an adequate understanding of the macroeconomic aspects that mars so many accounts of the robot and AI revolution and leads their authors to false conclusions.

Moreover, this book is, after all, about the *economic* consequences of robots and AI. These consequences have more to do with economics than they have to do with the intricacies of what robots and AI can do. Accordingly, it is entirely appropriate that Part I of the book is concerned with the macroeconomy. Even so, Chapter 2 does deal directly with robots and AI by putting current and prospective developments into perspective. Among other things, it discusses the extent to which robots and AI have implications fundamentally different from other technological developments that have occurred over the last 200 years.

But before this, Chapter 1 gives an account of our economic past, including those last 200 years. You may think that beginning a book about the future with this historical focus is quixotic. Yet it isn't. An understanding of our economic history is profoundly important for getting the present revolution in context. It is a tale rich in interest, full of surprises and brimful of relevance for the issues at hand. It is central to the debates about robots and AI to understand to what extent current and future technological advances are similar to what went before and to what extent they are fundamentally different.

Chapter 3 discusses the macroeconomic consequences of the advances of robots and AI. Will they lead to a recession or even, as some analysts allege, to a depression? Will the robot and AI revolution lead to a major reduction in employment opportunities? Will it be anti-inflationary? And what will it do for the economic growth rate and the growth of productivity and living standards? Moreover, in such a world, what will happen to interest rates and the various asset types in which we invest our money?

Part II is where the patience of readers who wish to get straight to the detailed shape of the future is rewarded. It is devoted to the consequences of the robot and AI revolution for the world of work and business, starting in Chapter 4 with a discussion of the rewards from work and the human need for it, balanced against the draw of increased opportunities for leisure. In it I reveal my vision of the likely future division between work, rest and play.

Chapter 5 describes the likely future shape of the labor market of the future: the types of job that will disappear, the types that will remain largely unaltered, the types that will survive but will be radically transformed, and the types that might appear from nowhere. And Chapter 6 identifies the likely winners and losers from these changes – not just groups of individuals but also regions and countries.

Part III is devoted to policy. Given the prospective changes described in the preceding chapters, what should governments do? Should they seek to encourage developments in AI or to restrain them, through taxation, regulation, or legal changes? This is the subject matter for Chapter 7.

Next, in Chapter 8, comes education. Given the prospective changes brought about by robots and AI, surely we cannot continue to educate children and university students in exactly the same way as we have up to now, as though nothing had happened. But how should they be educated? Will we need fewer teachers or more?

What subjects should be taught? And what is the appropriate role of the state in bringing about the necessary changes?

Chapter 9 deals with one of the most controversial issues of all, namely the idea that, because we may face a future in which extraordinary productive largesse coincides with widespread poverty, there needs to be a substantial redistribution of income and, perhaps, wealth. If society made this choice, could it achieve the desired result by reforming the current system for redistribution? Or should we adopt the radical suggestion that society should provide a guaranteed minimum income for all? This idea has been embraced by many influential thinkers across the political spectrum. But does it make sense? Is it affordable and what consequences would it have for the incentive to work and for the shape of society?

In the Conclusion I draw together the outcome of the discussions and analysis of the previous nine chapters and present what I think the main lessons are for individuals, companies, and governments. But this isn't quite the end – in more ways than one. As advertised above, the Epilogue launches into even more controversial territory, namely how the world might look if and when we experience what the AI experts term "the Singularity," when AI becomes more intelligent than humans and takes over the world – and/or AI and humans fuse together.

But the place to start on this adventure that may take us into the far future, and beyond, is surely with an understanding of how we got to where we are.

PART I

Man and machines: past, present, and future

1
The ascent of man

"Productivity growth isn't everything; but in the long run it's almost everything."

Paul Krugman[1]

"The past 250 years could turn out to be a unique episode in human history."

Robert Gordon[2]

If there is one event in our economic history that should be counted as a "singularity" it is surely the Industrial Revolution. Like all things historical that one learned about at school, the Industrial Revolution is more complicated than how it was presented back then. For a start, you could quite reasonably say that it wasn't a revolution. After all, it wasn't a single event but rather a process, starting in Great Britain in the late eighteenth century and drawn out over many decades.

And you could also say that it wasn't exclusively, or perhaps even primarily, industrial. There were certainly great advances in manufacturing, but there were also great advances in agriculture, commerce, and finance. Moreover, what made the Industrial Revolution possible – and what made it happen in Britain – was less to do with material factors, such as the availability of coal and water power that we had drilled into us at school, and more to do with the political and institutional changes that had happened over the previous century.

But no matter. Whatever name you want to give it, it was momentous. Before the Industrial Revolution there was next to no economic progress. After it, there was nothing but.

This is a simplification, of course. There was some growth in per capita output and incomes beforehand, including in both the USA and Britain in the seventeenth and eighteenth centuries – although the pace of advance was minimal compared to what came later.

Nor is it quite right to see economic progress as continuing relentlessly after the Industrial Revolution. As I will show in a moment, there have been some notable interruptions. Moreover, it took decades for there to be any increase in real living standards for ordinary people. [3]

These various quibbles and qualifications have led some economic historians to question whether we should dispense altogether with the idea of the "Industrial Revolution." Yet this would surely take revisionism too far. Rather like those historians who claim that, in contrast to their fearsome reputation, the Vikings were really nice, civilized, decent people, if not actually cuddly, in seeking to correct the simplicities of an established view, they have veered off too much in the other direction. The Vikings really were pretty scary, and the Industrial Revolution really was momentous.

One of the key characteristics of the post-Industrial Revolution world that marks it out as different from everything that came before is that, starting during the Victorian age in England, it came to be widely *believed* that the human condition would continue to get better and better, inevitably and inexorably. As the historian Ian Morris has put it, the Industrial Revolution "made mockery of all the drama of the world's earlier history."[4]

From ancient to modern times

This message about the significance of the Industrial Revolution is borne out by Figure 1. It shows what has happened to per capita GDP from 2000 BCE to the present. Believe it or not! Clearly, the early data shown in the chart are pretty dodgy. They should be regarded as indicative at best. In fact, more than that. You can ignore the absolute numbers marked on the chart's axes. They have no meaning. It is the relativities that should command attention. Per capita GDP in each year is compared with how it was in the year 1800. (In other words, all figures are indexed, with the year 1800 set equal to 100.)

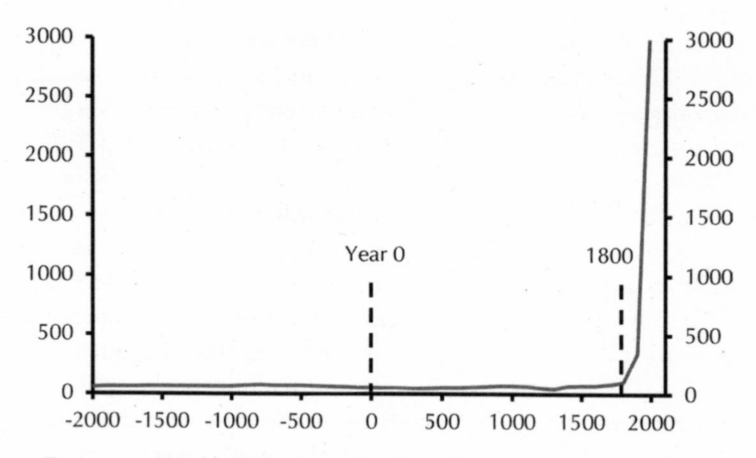

FIGURE 1 World GDP per capita from 2000 BCE to the present day (the year 1800 = 100)

Source: DELONG, Capital Economics

As the chart shows, there was effectively no change in per capita GDP from 2000 BCE to the birth of Christ, marked on the chart as Year 0. From then to 1800 per capita GDP doubled. This may not sound that bad but bear in mind that it took us 1,800 years to achieve this result! In regard to the rate of increase from one year to the next, it amounts to next to nothing. (That is why you can hardly make out any rise on the chart.) Moreover, the improvement was heavily concentrated in the later years.[5]

But after the Industrial Revolution things were utterly different. The chart clearly shows the lift-off. In 1900 per capita GDP was almost three and a half times the 1800 level. And in 2000 it was over thirty times the 1800 level.[6] The Industrial Revolution really was revolutionary. It provides the essential yardstick against which we must measure and assess the advent of robots and AI.[7]

In what follows, I trace out and discuss the major features of our economic history from ancient times, through the Industrial Revolution to the present. But I hope that readers will understand that, compared to my more detailed discussion of recent decades, coverage of earlier centuries is thinner and I breeze through time much more quickly. This is both because we have much less information about our distant history and because, as we contemplate the potential economic effects of robots and AI, ancient times are of less interest and relevance than recent decades.

Ancient puzzles

Right at the heart of the Industrial Revolution was technological change.[8] Yet there were some notable milestones in technological development well before the Industrial Revolution. Indeed, going a fair way back into history there were dramatic advances such as the domestication of animals, the plantation of crops, and the invention of the wheel. But they don't show up clearly in our record of world GDP per head. Believe it or not, this "record," or rather the economist Brad DeLong's heroic efforts to construct one, goes back to one million years BCE. (There is no point in extending Figure 1 this far back because all you would see is a practically flat line, and this would obscure the significance of what happened over the last 200 years.)

Now, admittedly, the absence of much recorded economic growth in earlier periods could simply be because our economic statistics are hopelessly inadequate. They certainly are poor and patchy. But we don't have only this inadequate data to go on. The evidence from art, archaeology, and such written accounts as we have, all point to the conclusion that the economic fundamentals of life did not change much across the centuries, at least once mankind abandoned nomadism for a settled life.

Why did these earlier, apparently revolutionary technological developments, referred to above, not bring an economic leap forward? The answer may shine a light on some of the key issues about economic growth that haunt us today, and in the process pose important questions about the robot and AI revolution.

I am sorry to say, though, that there isn't one clear and settled answer to this important historical question. Rather, four possible explanations suggest themselves. I am going to give you all four without coming to a verdict as to which explanation is the most cogent. We can leave that to the economic historians to scrap over. In any case, the truth may well be a mixture of all four. What's more, each of these possible explanations has resonance for the subject of this enquiry, namely the economic impact of robots and AI.

The first seems prosaic, but it is nonetheless important. Momentous developments such as the First Agricultural Revolution, involving the domestication of animals and the plantation of crops, which can be said to have begun about 10,000 BCE, were stretched out over a very long time. Accordingly, even if the cumulative effect once the process was complete was indeed momentous, the changes to average output and living standards did not amount to much on a year-by-year basis.[9]

The second possible explanation is structural and distributional. For a technological improvement in one sector (e.g., agriculture) to result in much increased productivity for the economy overall, the labor released in the rapidly improving sector has to be capable of being employed productively in other parts of the economy. But as the First Agricultural Revolution took hold, there were effectively no other forms of productive employment. Hence the proliferation of temple attendants, pyramid builders, and domestic servants. The anthropologist James Scott has suggested that after the First Agricultural Revolution average living standards for the mass of the population actually declined.[10] Nor was there anything in the new agrarian economy, with its lopsided income and wealth distribution, that favored further technological developments.

From technology to prosperity

The third possible explanation is that technological advance alone is not enough to deliver economic progress. You have to have the resources available to devote to new methods and to make the tools or equipment in which technological progress is usually embodied. Accordingly, growth requires the forgoing of current consumption in order to devote resources to provision for the future. Human nature being what it is, and the demands for immediate gratification being so pressing, this is easier said than done.

Unfortunately, the sketchiness of our data on the distant past again precludes us from conclusively establishing the truth on this matter.

But it seems likely that ancient societies were unable to generate much of a surplus of income over consumption that could be devoted to the accumulation of capital. And we also have to take account of the destruction of capital in the various wars and conflicts to which the ancient world was prone. So the net accumulation of capital was probably nugatory.

Such surpluses as were generated from ordinary activities seem to have gone predominantly into supporting the existence of non-productive parts of society, such as a priestly caste, or the construction of tombs and monuments. In ancient Egypt, heaven knows what proportion of GDP was devoted to the construction of the pyramids or, in medieval Europe, to the erection of the splendid, and splendidly extravagant, cathedrals that soared above the seas of poverty all around. It is wonderful that these remarkable buildings are there for us to enjoy today. But they didn't exactly do much for the living standards of the people who were awed by them when they were built – nor for the rate of technological progress, either then or subsequently.

The demographic factor

The fourth reason that technological progress did not automatically lead to increased living standards is that the population rose to soak up whatever advantages were gained in production. The evidence suggests that for the world as a whole there was annual average growth of about 0.3 percent in the sixteenth century, but population growth averaged 0.2 percent per annum, leaving the growth in GDP per capita at a mere 0.1 percent, or next to nothing. Similarly, in the eighteenth century, just before the Industrial Revolution, global growth appears to have averaged about 0.5 percent, but again just about all of this was matched by an increase in population, meaning that the growth in real GDP per head was negligible.[11]

Admittedly, the linkages here are not straightforward. After all, an increased number of people was not an unmitigated disaster, imposing a burden on society, as is so often, erroneously, assumed. On the contrary. More people meant more workers, and that would tend to increase overall production. But applying more workers to a given amount of capital and land would tend to produce lower average output. (Economists know this as the law of diminishing returns.) Moreover, a higher rate of population growth would mean a higher ratio of nonproductive children to productive adults. (Mind you, just as in many poor societies today, stringent efforts were made to make children to some degree productive from an early age.)

The constraints on living standards imposed by rising population were the central element in the theory propounded by the Reverend Thomas Malthus, who was both a minister of the Church and one of the early economists. These days his rank pessimism has been completely discredited. And rightly so. He really did give economics and economists a bad name. Writing in England in 1798, he said:

> *The power of population is so superior to the power of the earth to produce subsistence for man, that premature death must in some shape or other visit the human race. The vices of mankind are active and able ministers of depopulation. They are the precursors in the great army of destruction; and often finish the dreadful work themselves. But should they fail in this war of extermination, sickly seasons, epidemics, pestilence, and plague advance in terrific array, and sweep off their thousands and ten thousands. Should success be still incomplete, gigantic inevitable famine stalks in the rear, and with one mighty blow levels the population with the food of the world.*[12]

In one of the rare forays of an economist into the realms of carnal desire, he warned that the "passion between the sexes," if left unregulated, would result in misery and vice. He urged that the "consequences of our natural passions" be frequently brought to "the test of utility."[13]

Here lies a lesson for all those authors, both techie and economic, who currently wax lyrical on the horrors that human beings face in the robot-dominated future. You are entitled to be that gloomy if you like, but if you are, and wish to preserve your reputation, you had better be right.

Poor old Malthus. If ever an economist got things comprehensively wrong, it was him. Over the last 200 years – although not quite as soon as the ink was dry on his writings – per capita GDP and living standards have shown a dramatic rise. From 1798, the year Malthus published his gloomy tome, to today, the cumulative increase in real per capita GDP in the UK has been over 1,300 percent. And the increase in living standards (on which we don't have full data) would not be much different. (Admittedly, this improvement did not occur initially, as I will discuss in a moment.) And all the while, the population has been rising relentlessly. It would surely not be too unkind to say that nothing remotely like this was envisaged by the Reverend Malthus. Indeed, to put it baldly, these facts completely contradict his thesis.

Malthus failed to foresee two major elements of economic progress. First, he not only underestimated the scope for technological progress in general, but he appears to have misread entirely the scope for technological progress in the production of food in particular. In the nineteenth and twentieth centuries we managed to sharply increase the output of food, not just by taking into cultivation new land in the Americas and elsewhere, but also because, through advances in the techniques of food production, we were able to increase the output from a given amount of land.

Second, through the use of various methods of birth control, in the last century the birth rate fell back. This meant that, although population continued to rise, it did not increase anything like fast enough to prevent living standards from increasing.

This does not mean, though, that Malthus was wrong about the rest of human history. Indeed, up to the Industrial Revolution he seems to have been broadly right. Anyway, we shouldn't feel too sorry for him. Posthumously he earned one of the greatest consolation prizes going: Charles Darwin credited him as the inspiration for his theory of evolution through natural selection.[14]

Fluctuations and losers from change

The account I gave at the beginning of this chapter makes it sound as though the post-Industrial Revolution world has been one long upward trajectory. That is indeed the impression given by Figure 1. And it is a very good first approximation of what happened. But it is not the whole truth. Once the engine of economic progress got going, there was anything but a smooth and even process of rising living standards for everyone. Indeed, the early decades of the nineteenth century constituted a prolonged period during which wage growth lagged productivity growth, and living standards were squeezed. This period is known as "the Engels pause," after Friedrich Engels, Karl Marx's collaborator and benefactor, who wrote about it in *The Communist Manifesto*, published in 1848.[15]

The historian Yuval Noah Harari suggests that in 1850 (i.e., before the consequences of the Industrial Revolution had started to elevate general living standards), "the life of the average person was not better – and might actually have been worse – than the lives of archaic hunter-gatherers."[16] Similarly, the economic historian Robert Allen has argued that it was only after 1870 that European real wages rose decisively above medieval levels, with Britain leading the way. Indeed, for many parts of Europe, he says that it is difficult to argue that the standard of living in 1900 was notably higher than it had been in the sixteenth century.[17]

Moreover, the economy as a whole was subject to marked fluctuations. For particular trades, regions, and countries, these fluctuations were even greater. Of course, well before the Industrial Revolution there were also fluctuations in fortunes. The Bible talks of seven years of plenty and seven of want. Such fluctuations were usually caused by variations in the harvest, but disease, natural disaster, war, and civil disturbance also played a part.

These things continued up to and after the Industrial Revolution. But over and above these sources of vagaries in fortune, the new market exchange economy that became dominant from the eighteenth century onward was also subject to swings of *aggregate demand*, or spending power. As a result, there were periods of substantial unemployment during which, even if workers' skills were not redundant, there was inadequate demand for them. This feature of the money-exchange economy was most evident during the Great Depression of the 1930s, marked by mass unemployment in many countries. (There's more about this in Chapter 3.)

Furthermore, the technological "progress" that underpinned the Industrial Revolution undermined the livelihoods of many individuals and groups. This was not an unfortunate, incidental extra; it was intrinsic to the very process of economic growth, which relied upon old skills and occupations becoming redundant and new ones taking their place. The great Austrian American economist Joseph Schumpeter called this process "creative destruction."

Admittedly, even before the Industrial Revolution, there were also some cases of technological redundancy. For instance, the Venetian shipwrights, who for centuries had made their living out of constructing galleys and ships with fixed sails to ply their trade across the Mediterranean, eventually faced redundancy when oceangoing ships with adjustable sails came to dominate international trade.

Nor was it open to them simply to learn how to make these different types of vessel. The dominant trade routes changed as well. Asian trade with Europe no longer passed across land to the Eastern Mediterranean, and hence to Venice, but rather went by sea around Africa. And soon the transatlantic trade, conducted by the Atlantic-facing countries – Spain, Portugal, France, Holland, and England – dramatically rose in importance, too. So they were snookered.

Yet, with the Industrial Revolution and the mass movement of people from the land to cities, the chances of unemployment and/or impoverishment as a result of technological change and/or a slump in demand dramatically increased. Most people were now specialized in a particular trade and depended for their food, clothing, and shelter on being able to sell the fruits of their labor. This made them vulnerable because the particular skill that they acquired, or even their bare unskilled, brute labor, could become redundant thanks to technological progress (or some new commercial development).

Throughout history technological advances have been resisted by those whose livelihoods have been undermined by them. Understandably. In the fifteenth century, well before the Industrial Revolution, textile workers in Holland attacked textile looms by throwing their clogs (wooden shoes) into them. These shoes were called *sabots*. This may well be the origin of the word "saboteur."

It is scarcely surprising, then, that the early years of the Industrial Revolution witnessed resistance to new methods from groups of workers who saw their livelihoods threatened. In some cases in early nineteenth-century England, groups of workers gathered together to smash the machines that they considered to be threatening their wellbeing. They became known as "Luddites," after one Ned Ludd, possibly born Edward Ludlam, who is supposed to have smashed two knitting frames in 1779. Echoes of these attitudes and behavior continued throughout the nineteenth century and indeed continue even to the present. To this day, people who oppose technological developments are often branded "Luddites."

Nor was opposition to technological progress restricted only to those directly disadvantaged by it. In the third edition of his *Principles of Political Economy and Taxation*, published in 1821, the great economist David Ricardo added a new chapter, "On Machinery."[18] In it he said: "I am convinced that the substitution of machinery for human labour is often very injurious to the interests of the class of labourers." There have been frequent echoes of this thinking throughout the subsequent 200 years.

New jobs for old

Great economist though Ricardo was, as things turned out, his pessimism was unjustified. Although many individual workers experienced loss of employment or diminished incomes as a result of technological advances, this is not the story of the economy as a whole. Jobs lost in some sectors or occupations were replaced by new jobs created in others. The most dramatic impact of mechanization on jobs was probably in agriculture. As late as 1900, agriculture accounted for 40 percent of employment in the USA. In 1950 it accounted for 12 percent, and the figure today is 2 percent. In the UK the corresponding figures are 9 percent in 1900, 5 percent in 1950, and 1 percent today.

The reduction in the relative importance of agriculture and the decline of employment opportunities in that activity turned out to be good news for humans. But, at least in terms of numbers, it was not good news for horses. In 1915 the horse population of the USA was some 26 million. This has been referred to as the time of "peak horse." Today the horse population stands at about 10 million. And the change isn't only a matter of numbers. In 1915 nearly all the horses were a vital part of the productive process. Today they are almost all used for some sort of leisure activity.[19]

Some pessimistic commentators have seized on the analogy with horses to suggest that the current phase in economic development can be regarded as "peak human." If they are right, I suppose human existence, too, would be driven by leisure activities, just like the remaining horses. (Whether this vision is going to turn into reality is the subject of succeeding chapters.)

After the share of agriculture in the economy contracted in favor of manufacturing, something similar subsequently happened with manufacturing as its share fell in favor of services. In the United Kingdom in 1901, manufacturing accounted for almost 40 percent of total employment, with services accounting for only slightly more. Now, manufacturing is down to 8 percent of all jobs while services account for 83 percent.[20]

Moreover, within the broad categories of employment, such as agriculture, manufacturing, and services, it has been a similar story of existing jobs destroyed and new jobs created. In the 100 years to 1971, the number of people employed as telephone and telegraph operators rose by a factor of forty. Since then, the advent of automated switchboards, as well as the internet and mobile technology, has caused such employment to shrink dramatically.

By contrast, over the last 35 years, the number of information technology managers in the United Kingdom has risen by a factor of more than 6 and the number of programers and software development professionals has risen by a factor of almost 3.[21]

Productivity-enhancing changes have sometimes expanded employment even in the industries undergoing the improvement. Henry Ford's introduction of assembly line production methods is a case in point. In 1909 it took over 400 working hours to produce a car. Two decades later this had fallen to less than 50 hours. Yet employment in automotive manufacturing took off. Much greater efficiency in production translated into lower prices, and this, combined with other factors, led to a greatly expanded demand for cars.

The more normal experience, though, has been for employment in the industries enjoying rapid productivity growth to contract. But the fall in the prices of the goods produced by those industries has raised the real incomes of consumers and led to increased demand, not just for the goods whose prices have fallen, but for a wide variety of other goods and services, leading to increased employment in the industries producing these other goods and services.

The result is that those critics – and there were lots of them – who saw technological change as destroying large numbers of jobs and even threatening employment overall have been proved wrong. There has been no tendency for overall employment to diminish. Quite the opposite. Indeed, not only has overall employment kept rising but in the UK, as well as some other developed countries, the ratio of employed people to the overall population has recently been rising, too.

And wages and salaries have risen substantially as well. From 1750 to the present, although there were some pauses and reversals, and although the picture may have changed a bit recently, the share of wages and salaries in national income has been broadly constant. This means that the benefits of productivity growth were shared more or less equally between the providers of labor (the workers) and the owners of capital (the capitalists). The relentless rise in average real wages and salaries made possible a corresponding improvement in the average standard of living.[22]

Still, we all know about averages. If people were to overcome the problems that technological change thrust upon them, they had to adapt themselves – to learn new skills and/or to shift location. Many individuals did manage to do this, but some did not. Accordingly, during this period of "progress," large numbers of people underwent immense suffering.

Technology and the engine of growth

Before we leave the Industrial Revolution and move on to more recent times, we need to get technological change into perspective. Economic history is full of inventions. And economic history books about the Industrial Revolution are fuller still. All steam engines and "spinning jennies." This isn't wrong, but it is partial, and it can be misleading. It is true that productivity is the key to economic growth – certainly to the growth of output per capita, which is the ultimate determinant of living standards.

But there is more to productivity growth than inventions and technology. A society can enjoy increased living standards over time if it devotes a proportion of its output to real investment, over and above the amount that is necessary to replace the stuff that has been worn out by age, continued use, or wartime destruction. Sustained net real investment will mean that the amount of capital that workers have to work with will rise over time and this will result in higher output per head, even if there is no technological progress.

Moreover, sometimes output per head can increase even without having the benefit of either new inventions or more capital. At the very basic level, over time human beings and their organizations (families, firms, and governments) learn to do things better, bit by bit, incrementally. (In the economics literature, this is known as "learning by doing.") This increases output and productivity.

Sometimes there can be great leaps forward as a result of trade and commerce. This can happen through internal barriers to trade being removed (as with the *Zollverein* among mid-nineteenth-century German states). Or it may occur as the result of the discovery and subsequent development of new land, such as happened with the Americas and Australia. In the early years of the development of both the Americas and Australia, commercial expansion took place without any assistance from new technology. (Admittedly, in the late nineteenth century it was greatly aided by the emergence of steamships and refrigeration.)

And then, quite separately and at first unconnected with international trade and commerce, every so often come the great inventions and technological leaps forward.

Different periods experience a different balance of these various factors. The Industrial Revolution saw a combination of all of them. And the benefits were interactive as the expansion of markets made possible by increased trade enabled the exploitation of economies of scale in production – just as Adam Smith argued would happen in *The Wealth of Nations*, published in 1776.

Economic development continued throughout the nineteenth century and into the twentieth. But twentieth-century economic performance did not at first show the full benefits because other factors intervened. The first half of the century was scarred by two world wars that destroyed capital on a massive scale and diverted resources into war production. And sandwiched between these two devastating conflicts was the Great Depression which saw a large loss of output in most countries in the developed world. In the USA, GDP fell by 30 percent and unemployment reached a peak of 25 percent of the workforce. Still, all things come to an end, including bad ones. And the end of the Second World War set the stage for something extraordinary.

The postwar boom

The period from the end of the Second World War to 1973 (the year of the first OPEC oil price hike) was one of the most remarkable in our history. Several of the sources of economic growth discussed above came together at the same time:

- After the destruction of the war, there was an urgent need for substantial postwar reconstruction. This was greatly assisted by the financial aid provided by the USA through the Marshall Plan.
- There was a hangover of inventions and developments from the 1930s and the war years that had not been put to full commercial use. This, and the drip-feed of new advances, led to a steady stream of technological improvements.
- The economy was run at a very high level of aggregate demand and high employment.
- Thanks to the above factors, and low interest rates, investment was high.
- The international trading system was gradually liberalized, with the result that international trade took off, thereby realizing the gains from specialization that economists from Adam Smith and David Ricardo onward had lauded.

Accordingly, it is unsurprising that the years after the Second World War were a boom time for most of the developed world. From 1950–1973 most countries in the West enjoyed the greatest period of economic expansion in their history. During these years overall world GDP grew by 4.8 percent per annum on average. Even after adjusting for the growth of the population of almost 2 percent per annum, the average growth of GDP per capita was 2.8 percent. This average includes many countries that were not growing by much at all. And, in the way of all averages, this means that some countries enjoyed much higher growth rates. Over this period the average annual growth of GDP per capita in West Germany was a staggering 5.6 percent.

Although most economists would see such growth rates as spectacular, to the untutored eye they may not seem very impressive at all. But compound interest is a wonderful thing. When compounded over 23 years, a growth rate of 2.8 percent produces an overall increase in GDP per capita of almost 90 percent. And for an annual growth rate of 5.6 percent, the rate recorded in West Germany, the total increase is 250 percent. No wonder that for the industrialized world as a whole economists generally refer to this period as the "Golden Age," and no wonder that Germans refer to this period as the *Wirtschaftswunder*, or economic miracle.

But good things come to an end as well. After this period, across almost all countries, growth fell back sharply. For about 15 years the world as a whole registered growth barely higher than what had been experienced from 1870 to 1913. The reasons for this sharp change continue to be disputed but it seems clear that the dramatic hikes in oil prices in 1973/74, and again in 1979/80, played a major part. Alongside this, the international monetary system fractured as the fixed exchange rate regime based on the dollar broke down, and inflation took off.

After a period of high inflation in the 1970s, central banks and governments then embarked on policies to bring inflation down again, involving sky-high interest rates. They eventually succeeded in putting a lid on inflation but at the cost of very high levels of unemployment. This is when the term "stagflation" became commonly used. A mood of pessimism set in about economic prospects.

Subsequently, economic growth picked up, at first centered on the developed economies but later driven by rapid growth in the emerging markets. Between 2001 and 2007 global growth averaged almost 4 percent, not quite as high as what had been experienced during the Golden Age but much better than what had been experienced since then. This gave the impression that the good times were back. In fact, these global aggregates were misleading. Excluding the extraordinarily high growth rates in the emerging markets, led by China, in most countries in the West growth rates were trending down.

These rapid growth rates in the emerging markets were not driven by some new, transformative technology. The emerging markets' spurt of rapid growth derived mostly from the dramatic gap between the levels of GDP per capita in the emerging markets and equivalent levels in the West. This allowed enormous scope for catch-up growth. All the emerging market countries had to do was to adopt the technologies already employed in the West and, hey presto, rapid growth would ensue. I say "all they had to do" but this risks downplaying the scale of the transformation required to close, or at least reduce, the gap with the West. This involved dramatic social and political changes, including the effective abandonment of red in tooth and claw communism in China.

Yet, as the emerging markets got closer to the leaders, the rate of growth made possible by catch-up alone started to fall back. And the huge boost to output and living standards from opening up to international trade, although it continues today, was a one-off transformation. You cannot keep on opening up.

Significantly, although it was helped by computers and the internet, the sources of this great surge of growth in the emerging markets, buoying the growth rate of the world as a whole, which took umpteen millions of people out of poverty, were essentially commercial and political, rather than technological.

And then came a series of events that shook the world.[23] As it happens, they weren't technological either. The roots of the Global Financial Crisis (GFC) of 2007–9, and hence of the subsequent Great Recession, lay in the world of finance, not technology. And, despite the involvement of so-called *financial* technology in what went wrong, at bottom the causes of the GFC lay in some abiding weaknesses in human nature, institutions, and public policy.[24]

The great slowdown

In the years since the GFC and the subsequent Great Recession, the measured rate of economic advance has been subdued.[25] This dramatic weakening in the pace of economic expansion has unleashed a flood of academic papers trying to explain it. Some economists have argued that this is "the new normal." Many of them have focused on the demand aspects of this slowdown. The distinguished American economist Larry Summers has revived the notion of "secular stagnation" that did the rounds in the 1930s and survived for a while after the war – until the prolonged postwar expansion dealt what seemed to be a killer blow. (It seems that economic ideas never really die; they merely shuffle offstage, apparently mortally wounded, only to return at a later date.)

The essential notion behind secular stagnation is that, for a variety of reasons, aggregate demand displays a persistent tendency to fall short of aggregate supply. The result is that to prevent this from causing widespread unemployment, the fiscal and monetary authorities have to resort to exceptional stimulatory policies such as large increases in budget deficits, near-zero interest rates, and huge programs of bond purchases paid for with newly created money.

Some, but by no means all, of the economists supporting this view emphasize an increase in inequality as the fundamental underlying cause of secular stagnation. (American Nobel laureate Joseph Stiglitz is a notable exponent of this view.) Other economists accept the correlation but argue that the causation runs in the opposite direction, that is to say, the weakening in the growth of productivity has increased inequality.[26] Others tend to attribute the subdued growth of aggregate demand primarily to the weakness of the banks, the overhang of debt and various regulations, and the various inhibitions and deterrents that have made companies more wary and cautious. This debate is going to run and run. Fortunately, we have no need to prejudge the outcome here. But we do have to discuss some of the other possible explanations for the economic slowdown.

The technological explanation

The demand-led explanation for weak economic performance can readily take on a supply-side aspect as well. For weak aggregate demand often includes, or at least leads on to, weak investment spending on plant, machines, buildings, and software. A reduction in such spending will restrain the growth of productivity because it reduces the amount of capital that workers have to work with.

But some economists provide an explanation for the current state of the world economy that is explicitly driven from the supply side. Indeed, it is essentially technological. That is to say, they argue that weak growth is at bottom due to the slow rate of technological progress.[27]

The most famous and the most cogent of these economists is the American scholar Robert Gordon.[28] He argues that the recent slowdown in global growth cannot be blamed simply on weakness of aggregate demand or slower growth of the working population. From what we can tell, the underlying growth of productivity appears to have slowed markedly.

Gordon says that we should see this in a long historical context. He argues that we should look upon the Industrial Revolution as a one-off event. There was little economic progress before it, and there may be next to no economic progress after it, or rather after the third and most recent of the *industrial revolutions* has played out. For he says that there have been three industrial revolutions so far. And we may now be about to experience a fourth, thanks to robots and AI.

Each of the past industrial revolutions was associated with key technological breakthroughs. The first industrial revolution, *the* Industrial Revolution, stretching roughly from 1750 to 1830, was associated with steam engines, cotton spinning, and the early development of railroads. (The economic impact of railroads, though, came decades later.) After this came the second industrial revolution, stretching roughly between 1870 and 1900, associated with the discovery of electricity, the internal combustion engine, and the telephone, followed by various complementary inventions such as numerous electrical consumer appliances, motor cars, and airplanes.

The third industrial revolution, which Gordon dates from about 1960, was associated with the computer, which led on to the World Wide Web and mobile phones.

Gordon argues that the recent slowdown in productivity growth is due to the fact that the third industrial revolution, whose closing stages are playing out today, was nothing like as significant as the other two. He has said: "Many of the inventions that replaced tedious and repetitive clerical work by computers happened a long time ago, in the 1970s and 1980s."

Meanwhile, more recent developments have been focused on improvements in entertainment and communication devices. Admittedly, these have provided opportunities for leisure and consumption, both at home and on the job. But they have not done much to increase output per head. In particular, they don't conform to the traditional pattern of technological advance that has underpinned economic progress over the last two centuries, namely the replacement of human labor by machines.

You would have thought that computerization most definitely did conform to this paradigm. Yet the Nobel Prize-winning economist Robert Solow famously remarked in 1987: "you can see the computer age everywhere but in the productivity statistics." [29] (Mind you, the pickup in US productivity in the late 1990s suggests that the gains from computers were real but, as with many other advances, delayed.) The American entrepreneur and venture capitalist, Peter Thiel, has put recent technological disappointment more pithily. He has said: "We wanted flying cars. Instead, we got 140 characters."

This contention of Robert Gordon's that technological progress has largely run its course is sensational. Think of it: the rapid development of the emerging markets slowing inexorably; overall economic progress essentially dribbling away to nothing; living standards barely rising at all; no prospect of the next generation being appreciably better off than the current one; a return to the situation and the outlook (if not the living standards) that pertained before the Industrial Revolution. There is no doubt that, if this is indeed our future, then it represents a dramatic transformation – with potentially serious political consequences. So, the stakes for all of us are very high. But is Gordon right?

Malthus all over again?

There are broadly four major counterarguments to the Gordon-led pessimism about current and future productivity growth. They are not necessarily contradictory or even competing, and I am not going to attempt to award marks to each. Again, I will leave that to others. Suffice it to say that I think they all have merit and that, not for the first time, the "true" answer would involve a mixture of all four.

The first counterargument is that we are simply not measuring productivity growth properly. Accordingly, the productivity slowdown may be simply a statistical illusion. As such, there may be no convincing basis for assuming that our extended future involves extremely slow growth of productivity, and hence living standards.

This seems all too plausible. After all, in anything but the simplest of economies, productivity growth is difficult to measure at the best of times. But we have just passed through the digital convulsion that has reshaped the way we live, and revolutionized industries from advertising and newspapers to banking.

Moreover, many of the new services that are available in this new digital world are provided free to the user, in contrast to the non-digital versions with which they are in competition, and which they often replace. For instance, take clicking on a YouTube video instead of going to the cinema. Given that for economists the value of something is closely linked to the price that it commands in the market, this is a particularly difficult problem. But just because some digital service does not have a price, this does not mean that its value is zero.

There are clear indications that the under-recording of GDP (and therefore of productivity) is significant. In the USA, the Bureau of Economic Analysis reports that the information sector's contribution to GDP is currently just under 4 percent, almost identical to its contribution in 1980, before the World Wide Web was invented, never mind everything else.[30] Can this be true?

Various attempts have been made to estimate the extent of the under-recording of output and hence productivity. In the UK, a study led by Sir Charles Bean, a former Deputy Governor of the Bank of England, concluded that over the years 2005–14, correcting for the under-recording of the digital economy could add between

0.35 percent and 0.65 percent per annum to the growth rate of GDP. This may not sound like much but if you add this to the measured growth of productivity you get a figure that, excluding the so-called Golden Age from 1950 to 1973, is not much different from the average recorded since 1800. And Sir Charles Bean's estimate may be under-egging the pudding.

This conclusion is backed up by Professor Martin Feldstein, who argues that the task of adjusting the GDP statistics to the various implicit price changes and quality improvements brought about by revolutionary technological change is "impossibly difficult."[31] Mind you, it is difficult partly because not all changes in the new digitized economy have brought improvements, as anyone who has tried calling their electricity or gas supplier can attest. So, it is a matter of weighing up the improvements against the aspects where things have got worse.

Time and tide

The second counterargument is that the sluggishness of the world economy is due to the lingering effects of the financial crisis of 2007–9 and the subsequent Great Recession. These caused corporate investment to fall back and businesses, including banks, to become more risk averse. The corollary is that, as the effects of the financial disaster fade, there is no reason why output growth and productivity growth should not return to something like normal. The strength of the world economy in 2017 and 2018 provides some support for this view.

The third counterargument is that the digital revolution needs time to play out. Economic history is full of relevant examples. The machine at the heart of the Industrial Revolution, the steam engine, was invented by Thomas Newcomen in 1712 to pump water out of flooded coal mines. It was more than 50 years later that James Watt, while repairing a Newcomen engine, developed it to produce more power, thereby enabling it to be deployed more widely.

Moreover, the distinguished historian Jared Diamond says that Newcomen himself based his engine on earlier prototypes developed by others.[32] Indeed, he claims that this is the normal pattern. We have been led to believe that the great inventors come up with a revolutionary new idea or piece of machinery out of nothing. In fact, they are usually building on foundations laid down by others.

What's more, although James Watt patented the first steam engine in 1769, it wasn't until about a hundred years later that the full impact on labor productivity was felt. Similarly, it seems that electricity didn't have a major impact on GDP growth in the USA until about a half-century after the first generating stations were built.

Whatever the truth of these three counterarguments, the optimists have a fourth one up their sleeve. Far from running out of the capacity for radical technological changes, they say, we are on the brink of new developments that promise to bring rapid advances. As the economist Paul Romer puts it:

> *Economic growth occurs whenever people take resources and rearrange them in ways that make them more valuable ... every generation has perceived the limits to growth that finite resources and undesirable side effects would pose if no new ... ideas were discovered. And every generation has underestimated the potential for finding new ... ideas. We consistently fail to grasp how many ideas remain to be discovered ... Possibilities do not merely add up; they multiply.*[33]

The first of the new "ideas" is biotechnology, which may make a major contribution to agricultural production, food processing, and protection of the environment. And developments in medical science promise both major improvements in the quality of life and a much-extended lifespan. Meanwhile, nanotechnology and 3D printing also hold out the prospect of increases in productivity in many traditional areas of manufacturing. And now, on top of this, we have robots and AI.

Back to the future

This lightning tour of our economic history has brought out several key points:

- Continued improvement in living standards has not been our lot since the beginning of time. Indeed, for most of our history there has been next to no improvement from year to year.

- Some advances in technology that seem to have been dramatic had little effect on the growth of GDP or living standards from year to year. In other words, the pace of change was not transformative.
- By contrast, the Industrial Revolution stands out as a mega "event" in our history, marking the point after which people, on average, became steadily better off.
- Yet, from the first, there were losers in this process of "improvement." Lots of people were absolutely worse off as a result of the changes. It took decades for everyone to be better off.
- Over time, jobs and livelihoods destroyed were more than made up for by new jobs created, often in spheres that could not have been imagined.
- Although it has been far from the only source of improvement, the most important reason for things getting continually better has been the progress of technology. This has made possible both the production of more output with the same or fewer inputs and the development of new products and services altogether.
- The greatest advances in productivity and living standards have occurred when technological advances have coincided with commercial, political, and social changes that have encouraged their full exploitation.
- In fact, without the right social, political, and commercial environment, on its own, technology has either not been able to advance, or, if it has, it has failed to deliver the goods.
- According to the official figures, the rate of growth of productivity has recently slowed considerably, possibly marking the gradual exhaustion of the scope for technical progress, and an increase in living standards.
- But it is unlikely that the official data are telling the whole story. The underlying improvement in human welfare has probably been much greater than the official data suggest.

And now we come to robots and AI. Just as with steam power and electricity before it, the initial development of AI has already affected several niche areas dramatically, but it now looks set to affect just about all parts of the economy, and indeed all aspects of our lives.

Moreover, in marked contrast to Gordon's characterization of the essence of the third industrial revolution, what happens here, in what is widely referred to as the fourth industrial revolution, most definitely does promise to conform to the normal pattern of technological advance throughout our industrial history, namely the replacement of human labor by machines. This promises to be a return to normal with a vengeance. If this happens on any scale, then the robotics and AI revolution will surely not be of little economic importance, as Gordon alleges for computers and the digital revolution. Indeed, quite the opposite.

So, just as many economists have started to become pessimistic about the capacity for economic progress and rising living standards, along has come a new "revolution" which promises to deliver just what they were beginning to despair of. But is it all that it is cracked up to be? And, if it is, so that the post-Industrial Revolution engine of progress is back in business, will the overall effects be the same as those that dominated the nineteenth and twentieth centuries? That is to say, will the "creative destruction" that powered previous advances again ensure that there are new jobs to take the place of the old? Or will it this time be a case of destructive destruction?

2
Could this time be different?

"The pace of change has never been this fast, and yet it will never be this slow again."

Justin Trudeau, 2018[1]

"Machines will be capable, within twenty years, of doing any work a man can do."

Herbert Simon, 1965[2]

The legendary investor Sir John Templeton once said that the four most expensive words in investing are: "This time it's different."[3] He was referring to financial investment, and he was commenting on the long line of arguments that have, from time to time, been adduced to justify ridiculously high asset prices. And it is never different. Bubbles will be bubbles. When they burst, all the prior special pleadings in favor of the assets in question look like empty financial propaganda. Yet this seems to be forgotten when the next investment fad appears, and the pattern repeats itself. Sir John's wisdom found handsome vindication in the collapse of the dot-com bubble in 2000–2, and again shortly afterward in the collapse of the American subprime mortgage market in 2007–10.

Economic, as opposed to financial, history is a different kettle of fish. Nevertheless, there is a parallel. Ever since the Industrial Revolution there have been plenty of people who have seen technological change in a negative light, arguing that it would impoverish ordinary people and bring unemployment on a massive scale. The naysayers have always been proved wrong. In his bestselling book *The End of Work*, economist Jeremy Rifkin envisaged a future in which automation and information technology boosted productivity, but a great mass of workers were left on the scrapheap and would not participate in society's success. The year of publication was 1995.[4]

But are the changes taking place now, thanks to robots and AI, essentially a continuation of what we have seen since the Industrial Revolution? Or are they something quite different?

As we saw in the previous chapter, the key characteristics of technological change since the Industrial Revolution have been a relentless increase in productivity that has underpinned a dramatic increase in living standards, and the replacement of many jobs lost because of these technological advances with new jobs created.

There are essentially two broad strands to the view that this time it really is different. The first is that developments with regard to robots and AI are really a non-revolution. In essence, this contention amounts to an extension of Robert Gordon's criticisms of the communications revolution that we encountered in the previous chapter. The critics say that there is a great deal of sound and fury about this subject, but it signifies not very much at all. In essence, they say that this "revolution" is different from everything that has happened since the Industrial Revolution because the engine of economic progress has ground to a halt. We are left with just illusions and hype.

The second strand of negativity about AI amounts to exactly the opposite. The AI revolution *is* quite unlike anything that has been seen before in regard to speed and scope. More importantly, because it essentially undermines the demand for just about all human labor, this revolution puts an end to the creation of new jobs to offset the loss of old ones. It is not the end of economic "progress," but it may well be the end of human advance.

We need to look at each of these strands of criticism in turn before coming to an overall conclusion.

Truly momentous

So, is the AI revolution all it is cracked up to be? It is easy to play down its significance, but we should beware. This wouldn't be the first time that a technological shift was underestimated. Indeed, underestimation of the power of new technology has a distinguished history. In 1943, Thomas J. Watson, a past Chairman of IBM, is supposed to have declared that "there is a world market for about five computers." In 1949, Popular Mechanics, thought to be a reputable journal, said that "computers in the future may weigh no more than 1½ tons."[5]

Fast-forwarding, from the very beginning, the development of the internet met with outright skepticism. In late 1996, *Time* magazine explained why it would never be mainstream. The magazine said: "It was not designed for doing commerce, and it does not gracefully accommodate new arrivals." In February 1998, *Newsweek* ran the headline: "The Internet? Bah!"

Perhaps most strikingly, the author of the article was Cliff Stoll, an astrophysicist and network expert. He said that online shopping and online communities were an unrealistic fantasy that betrayed common sense. He said: "The truth is no online database is going to replace your newspaper." Stoll said that a digital world full of "interacting libraries, virtual communities, and electronic commerce" was "baloney."[6]

AI has met with a similar wall of skepticism – until recently. This skepticism has been fueled, among other things, by the fact that AI has been with us for some time – at least in theory – and it has not yet produced anything really dramatic. It grew out of digital computing, which was explored and developed at Bletchley Park in England during the Second World War, famously enabling the Nazis' Enigma code to be broken.

That feat is closely associated with the name of Alan Turing. Turing was also responsible for AI's early conceptual framework, publishing in 1950 the seminal paper "Computing Machinery and Intelligence." The subject was subsequently developed mainly in the USA and the UK. But it waxed and waned in both esteem and achievement.

Over the last decade, however, a number of key developments have come together to power AI forward:

- Enormous growth in computer processing power.
- Rapid growth in available data.
- The development of improved technologies, including advances in text and image, including facial, as well as voice, recognition.

- The development of "deep learning".
- The advent of algorithm-based decision-making.

So now AI seems close to its "James Watt moment." Just as the steam engine was in existence for some time before Watt developed it and it came to transform production, so AI, which has been on the scene for some time, is about to stage a leap forward.

Moreover, its impact is likely to be felt right across the economy. Some technological improvements are specific to particular sectors or narrow aspects of production and have only limited impact on the wider scheme of things. But every so often a development occurs that unleashes a technology with general applicability. We call these general-purpose technologies (GPTs). The steam engine was a GPT and AI promises to be one, too. That is why I have thought it appropriate to refer to the decades ahead of us as "the AI economy."

Robots and AI promise to have a major impact on productivity because in some areas they can replace humans altogether and in others they can greatly increase their output per hour or improve the quality and reliability of what they do. Perhaps most significantly of all, they can effectively provide many service sector workers, for example in healthcare and care for the elderly, with an effective tool, thereby promising to overcome the hitherto sluggish growth of productivity in the services sector.

What's more, the speed of advance in AI is impressive. AI visionaries often enthuse about the notion that developments are now proceeding *exponentially*, that is to say, the level in the next period is always greater than the level in the previous period by a certain multiple or percentage. For example, something that doubles every year is experiencing exponential growth. So is something that grows at 20 percent every year. The result is that a given rate of growth in percentage terms produces larger and larger absolute increases.

Whenever growth is exponential, because a large proportion of the cumulative change happens late in the process, it is easy, earlier in the process, to miss the significance of what is happening. The back-loaded nature of exponential growth goes some way to explaining why we sometimes overestimate the effect of technology in the short run and underestimate it in the long run.[7] This feature potentially spells wondrous things for the future capabilities of AI but it also reinforces the tendency of individuals, companies, and governments to do little or nothing to adapt – until it is too late.

The AI literature is full of spectacular examples of exponential growth, often couched in easy-to-read and engaging terms, that bring out clearly the contrast between the apparently sedate pace of change early on and the dramatic transformation later. Consider the following example from Calum Chace:

> *Imagine that you are in a football stadium ... which has been sealed to make it waterproof. The referee places a single drop of water in the middle of the pitch. One minute later she places two drops of water there. Another minute later, four drops, and so on. How long do you think it would take to fill the stadium with water? The answer is 49 minutes. But what is really surprising – and disturbing – is that after 45 minutes, the stadium is just 7 percent full. The people in the back seats are looking down and pointing out to each other that something significant is happening. Four minutes later they have drowned.*[8]

Exponential growth is at the root of what is known as "Moore's Law," which is generally taken to mean that the processing power of $1,000 worth of computer doubles every 18 months, or sometimes two years. Some analysts even suggest that there is exponential growth in the rate of exponential growth. Federico Pistono says that, whereas computer speed (per unit cost) doubled every three years between 1910 and 1950, and every two years between 1950 and 1966, it is now doubling every year. He claims that: "According to the available evidence, we can infer that this trend will continue for the foreseeable future, or at least another 30 years."[9]

Exponential growth rates abound in the world of computers and AI. On the basis of recent growth rates, it is possible to imagine robots outnumbering humans before very long. Dr. Ian Pearson, a British engineer, inventor, and novelist, says that robots will increase from about 57 million today to 9.4 billion within 30 years. This prediction is based on an assumption that Pearson says is "modest," namely that the robot population grows by 20 percent each year. (This is, of course, another example of exponential growth.)

It isn't only the speed and scope of advances in AI that impresses and appalls so many people and fills them with a sense of foreboding about the future. AI now threatens to replace people in activities that were recently thought to be exclusively human. It used to be thought that the game of chess would be beyond even the most capable computer. But in 1997 IBM's Deep Blue defeated the world's best player, Gary Kasparov. Deep Blue was able to evaluate between 100 and 200 million positions per second. Kasparov said: "I had played a lot of computers but had never experienced anything like this. I could feel – I could smell – a new kind of intelligence across the table."

In 2001 an IBM machine called Watson beat the best human players at the TV quiz game *Jeopardy!* In 2013 a DeepMind AI system taught itself to play Atari video games like *Breakout* and *Pong*, which involve hand–eye coordination. This was much more significant that it might have seemed. The AI system wasn't taught how to play video games, but rather how to learn to play the games.

Kevin Kelly thinks that AI has now made a decided leap forward, but its significance is still not fully appreciated. He writes:

> *Once a computer manages to perform a task better than humans, the task is widely dismissed as simple. People then say that the next task is really hard – until that task is accomplished by the computer and so on. Indeed, once a machine is able to accomplish a particular thing, we often stop referring to it as AI. Tesler's Theorem defines artificial intelligence as that which a machine cannot yet do.*[10]

And the category of things that a machine cannot do appears to be shrinking all the time. In 2016 an AI system developed by Google's DeepMind called AlphaGo beat Fan Hui, the European Champion at the board game Go. The system taught itself using a machine learning approach called "deep reinforcement learning." Two months later AlphaGo defeated the world champion four games to one. This result was regarded as especially impressive in Asia, where Go is much more popular than it is in Europe or America.

It is the internet that has impelled AI to much greater capability and intelligence. A key feature behind human beings' rise to dominion over the physical world was the development of exchange and specialization, which was a network effect. When the internet came along, connecting computers in a network, this transformed their capability.[11]

And then we have coming along soon what the British entrepreneur Kevin Ashton has called "The Internet of Things." IBM calls this the "Smarter Planet"; Cisco calls it the "Internet of Everything"; GE calls it "the Industrial Internet"; while the German government calls it "Industry 4.0." But all these terms refer to the same thing. The idea is simply that sensors, chips, and transmitters will be embedded in umpteen objects all around us.

The internet entrepreneur Marc Andreessen has said: "The end state is fairly obvious – every light, every doorknob will be connected to the internet."[12] And, of course, all this connectivity with the material world will be in just the form that can readily be analyzed and reacted to by robots and AI. But all such connected things will also be able to communicate directly with humans by "speaking" to them. Ironically, some AI enthusiasts suggest that this may bring our attitudes to "things" closer to how people viewed them in pretechnological times, and how they are viewed in parts of the non-Western world today, that is to say, as possessing some sort of spirit and identity.

So, whatever you think about the scope for human advancement in the AI economy – and in a moment I am going to give some major qualifications before coming to a preliminary overall assessment – what is happening in the AI world can surely not be dismissed lightly.

What is left for humanity?

Now we come to the second view about how the robot and AI revolution may be different from everything that has happened since the Industrial Revolution, namely the contention that this time there won't be new jobs to take the place of the old.

The technological progress that has lain at the heart of economic advance since the Industrial Revolution has followed a distinct pattern. At first, machines replaced human brawn, leaving humans to do more brainwork. More recently, particularly through the development of computers, machines have replaced some brainwork. (The Chinese word for computer translates as "electric brain.") But, at least until very recently, computers have replaced humans only where people have been engaged in repetitive activity, allowing them to do other, less repetitive, tasks. Yet, as discussed above, now AI is threatening to replace non-repetitive brainwork as well. Indeed, all brainwork.

Interestingly, the notion that machines would take over all work done by humans is not a new idea. The Greek philosopher Aristotle, writing in 350 BCE, suggested that, if automata (such as the ones allegedly made by the god Hephaestus) managed to become capable of doing any work that humans do, then all workers, including slaves, would be redundant.[13]

Where on earth in a world like this is the human being to find (and keep) his or her place? It seems as though, rather like the Saxon noble and opponent of the Normans, Hereward the Wake, being confined to a tiny marshy island near Ely by the forces of William the Conqueror, human beings are being forced to relinquish more and more territory. What is going to be left for them? Where is the final defense going to be? What happens when even the last marshy island is overcome? Is there going to be anything left that humans can do better than machines?

Moreover, unlike humans, robots and AI will not need to be paid or provided with benefits and pensions. So, given that robots are both more productive and cheaper, why would humans be employed at all?

If work for humans largely disappears, then, at the very least, this poses a profound problem of purpose. Since the struggle to earn a living has formed the basis of human existence since the beginning of time, what is to take its place? The future becomes one of endless leisure, involving the various opportunities and problems that I will discuss in Chapter 4. Suffice it to say that, if this transpired, endless "leisure," aka unemployment, would not necessarily represent nirvana for most people.

Quite apart from the potential problems of boredom and listlessness, will this be a life of plenty or will leisure be accompanied by poverty? For, if their labor is not wanted, then how will people earn any money? With no money earned, they will have nothing to spend. Without spending there will be no demand for the output of the robots. The result is that not only will humans be without work, but robots will be unemployed as well. (I discuss and analyze this, and other, dystopian views of our economic future in the next chapter.)

Disquiet about this prospect is not a possible future occurrence. It is with us in the here and now. According to a 2015 survey from Chapman University, Americans fear robots replacing humans in the workforce more than they fear death.[14]

Jobquake

How seriously should we take the threat to employment? Plenty of AI visionaries are very pessimistic. MIT Professor Max Tegmark is perhaps the most prominent. He takes issue with the idea that just as the last and present century have seen umpteen new job categories emerge from nowhere to take the place of jobs lost, so this pattern will be repeated in the next several years with regard to jobs lost to the advances of AI. He argues that the vast majority of today's occupations were already in existence a hundred years ago. Using data from the US Department of Labor, he discovered that 80 percent of the 2014 occupations already existed in 1914. Furthermore, the numbers of people employed in the 20 percent of new occupations were modest, with only 10 percent of the working population engaged in them.

The US economy is much bigger today than it was in 1914, and employs far more people, but the occupations are not new. He says: "and when we sort them by the number of jobs they provide, we have to go all the way down to twenty-first place in the list until we encounter a new occupation: software developer, who make up less than 1 percent of the U.S. jobs market."[15]

Following on in this vein, there are some pretty apocalyptic estimates of future unemployment from a number of analysts and forecasters. "The Millennium Project," established in 1996 by a combination of UN organizations and US academic bodies, produced a report entitled "2015–16 State of the Future," including a section on the future of work, based on a poll of 300 "experts" from different countries. Their verdict was that global unemployment would be "only" 16 percent in 2030, and still "only" 24 percent in 2050. So that's all right, then.[16]

A more credible analysis comes from McKinsey. It estimates that, if advanced societies switch rapidly to new technology, by 2030 as many as 700 million people could be displaced by robots. Even if the pace of adoption is more modest – which McKinsey expects – then about 375 million people, or 14 percent of all workers, would have to move jobs and retrain.

Mind you, a study from the OECD[17] has recently concluded that far fewer workers are at risk of being replaced by a robot than was previously thought. It concludes that in rich (OECD) countries, "only" about 14 percent of jobs are "highly automatable." Even so, the study concluded that the number of jobs at risk was still huge – about 66 million across 32 countries, including 13 million jobs in the USA alone.

A well-known and much quoted study by Carl Frey and Michael Osborne of Oxford University concluded that 47 percent of US jobs were vulnerable.[18] And the distinguished economic historian Joel Mokyr weighs in with those people who argue that the changes afoot today are on a different scale from previous periods of job loss.[19]

So, the AI revolution promises to be transformational. It will undermine the demand for human labor across a broad range of occupations, including many which, until recently, seemed immune to any threat from mechanization. AI reaches deeply into realms that were previously thought to be uniquely human. The impending change in the labor market is going to be real, substantial, and wide-ranging.

Hope for humans

But do the undoubted advances of robots and AI really spell the elimination – as opposed to the transformation of that many jobs? And if so, are there going to be new job opportunities to offset the loss of jobs in traditional activities? There are several reasons to believe that there will be and that the pessimists' fear of a future state of full unemployment will be confounded. AI will *not* spell Armageddon for employment.

For a start, despite the scary headlines, the McKinsey study referred to above stresses that it could be highly misleading to think of certain jobs for humans disappearing altogether. It estimates that fewer than 5 percent of jobs are entirely automatable but that most jobs include an element that could be done by machines. Indeed, it says that "about 60 percent of occupations could have 30 percent or more of their constituent activities automated." This tendency even extends to CEOs. McKinsey estimates that, using current technologies, activities taking up more than 20 percent of a CEO's time could be automated.

Strikingly, the McKinsey Report is not gloomy about overall employment prospects. It sees close parallels with what happened with the introduction of the personal computer. This, it says, "has enabled the creation of 15.8 million net new jobs in the USA since 1980, even after accounting for jobs displaced." Interestingly, many of the countries with the highest number of robots per worker (e.g., Singapore, Japan, Germany) also have some of the lowest unemployment rates.

According to McKinsey, the aspects of a job – any job – that are least susceptible to automation are creativity and sensing emotion. Admittedly, the firm says that in the US economy only 4 percent of work activities require creativity. But for "sensing emotion" the figure is 29 percent. These figures suggest that there is great scope for jobs for humans to be transformed in a way that majors in skills that humans uniquely possess, and in the process to create more job satisfaction.[20]

This word "creativity" needs interpretation. We are not talking only about the ability to do what Beethoven or Van Gogh did. At the most mundane level, all humans exhibit creativity in their everyday lives. Children exhibit this in abundance in the way they play. It includes the ability to innovate, and to develop new ways of doing old things.

Actually, to McKinsey's two key criteria for human "comparative advantage" I would add a third: the need for the exercise of common sense. Even the most "intelligent" AIs lack this facility. This will probably mean that even in job categories or areas of economic activity where machines will largely take over, there will still need to be a higher level of human oversight.[21]

Similar conclusions to McKinsey's have been reached by the OECD. The study mentioned earlier concluded that most jobs were difficult to automate because they required creativity, complex reasoning, the ability to carry out physical tasks in an unstructured work environment, and the ability to negotiate social relationships. The director of employment, labor, and social affairs at the OECD, Stefano Scarpetta, gives an interesting example that contrasts a car mechanic working on a production line in a huge plant with one working in an independent garage. The first is easy to automate whereas the latter is pretty hard to do so.

Using these criteria, even according to the AI visionary and employment pessimist Max Tegmark, there are several areas where, for the foreseeable future, AI will make no direct impact. These are the more creative activities, including journalism, advertising, all forms of persuasion and advocacy, art, music, and design. Mind you, he sees even these jobs disappearing in time and, what's more, a dearth of new jobs appearing to take the place of the old. I think that he is profoundly wrong about this, as I will show in a moment. (I will sketch out the likely shape of the future labor market in Chapter 5.)

A good deal of what separates human beings from AI can be summed up in the term "emotional intelligence." Actually, some work in AI research is trying to get robots to recognize the emotional state of humans with whom they are interacting and alter their behavior accordingly, even going so far as to appear to be empathetic. Good luck with that one! I suspect that robots pretending to have emotional reactions and an understanding of people would soon be regarded by humans as ridiculous.

The speed of advance

Perhaps surprisingly, there is a key source of comfort over the prospects for humans from the likely speed of advance in the capabilities of AI. I discussed above the power of Moore's Law, but in fact Moore's Law isn't a law at all. To give him his due, the AI enthusiast Calum Chace admits as much. He says: "There is no reason why the number of bits you can get onto a microchip will necessarily continue to increase exponentially. Similarly, for the amount of computing power that you can buy for $1,000. Indeed, there is good reason to suppose that both these growth rates will fall back substantially."[22]

The examples of exponential growth that pepper the AI literature are indeed impressive. They are meant to bowl you over with the combination of the simplicity of the process and the scale – and apparent inevitability – of the eventual outcome.

But everything hinges on the assumption of continued exponential growth. And, precisely because the effects of sustained exponential growth are so devastating, in the practical world it is seldom experienced for long periods. Often growth starts off slowly at first, then enters its exponential phase but then subsequently slows. This makes a sort of S-shape. Sometimes, the rate of growth trails off to a non-exponential path. This is the case when the number always increases by 10 every year (which means that the growth rate, expressed as the size of each annual increase as a percentage of the previous total, is continually falling).

In other theoretical cases, something can carry on growing and yet will never exceed a certain limit. This is the case of the frog that leaps half the distance between itself and the edge of the pond. It will get ever closer, but it will never reach the edge. In some other cases, something that was previously growing rapidly may come to a complete halt.

In the AI literature's frequent invocation of "exponential growth," there is a pattern that we will encounter throughout this book. An observed phenomenon is described as a "law" and conclusions that must "inevitably" follow are drawn. Often such assertions rest on flimsy, if not nonexistent, foundations. Disappointment and disillusion are the "inevitable" result.

Indeed, some good judges think that the pace of AI advances may recently have slowed down. This is the view of John Markoff, the Pulitzer Prize-winning *New York Times* journalist. He was struck by the disappointing performance of the robots entered into the DARPA Robotics Challenge in June 2015. Markoff claims that there has been no profound technological innovation since the invention of the smartphone in 2007.[23]

Technological underperformance

There are serious doubts about just how capable robots and AI will really be. Yes, there is plenty of evidence of some advances in technology, including in AI and robotics, being much faster than even the optimists have imagined.

But the picture of continual outperformance of our expectations is not the sum total of our experience with technological development in general or the development of AI in particular. Quite the contrary. One of my favorite examples is of a recent trip through passport control at Heathrow airport by the journalist Allison Pearson, late one evening. She was shocked by the fact that the automatic, that is, machine-driven, passport review points were closed and, as a result, there were huge queues for the few open, human-staffed, entry points. She asked an official why the machine entry points were closed and elicited the remarkable reply "staff shortages."

When she remonstrated that surely the whole point of these machines was to reduce the need for human labor she was told that, although this was true, the machines had a nasty habit of not working properly – and/or those pesky humans were not able to deal with them – with human passport control officers having to sort out the consequent mayhem. So, after normal working hours, given the shortage of staff, the machines had to be shut down.

I am sure that the glitches at passport control will in due course be sorted out. Indeed, I have recently experienced several completely glitch-free encounters with such machines in Madrid, Amsterdam – and even Heathrow. But the lesson – and this is far from being an isolated example – is that technological "improvement" often disappoints and that the full benefits take much longer to come through than the original developers and subsequent enthusiasts allege. Often the old systems, involving continued employment of human labor, have to run alongside the new, until the new systems are well and truly bedded down.

This point about technology in general applies just as much to AI in particular. In 1965, Herbert Simon said that "machines will be capable, within twenty years, of doing any work a man can do."[24] And in 1967, Marvin Minsky said: "Within a generation the problem of creating 'artificial intelligence' will substantially be solved."[25] Needless to say, these forecasts have proved to be hopelessly optimistic. These analysts are like latter-day Malthuses – only in reverse. (Mind you, their other achievements belie this comparison.)

Indeed, the history of AI research is one of alternate periods of feast and famine as the experience of success in some field leads to a flood of investment, followed by failure, leading to funding being cut, or even drying up altogether. Periods of the latter are referred to as "AI winters."

In fact, machines that matched humans in general intelligence have been anticipated since the invention of computers in the 1940s. At that time, and at all times subsequently, this development was thought likely to occur about 20 years into the future. But the expected arrival date has been receding at the rate of about one year per year. Accordingly, many futurists still put the arrival date of machine general intelligence equivalent to humans 20 years into the future.[26]

The Ridley Scott film *Blade Runner*, which depicted a dystopian future in which artificial life forms beat humans in both strength and intelligence, appeared in 1982. A sequel, *Blade Runner 2049*, appeared in 2017, again depicting a similar dystopian future.

Robots disappoint

Meanwhile, in the real world, the performance of robots has disappointed early hopes. You could be forgiven for thinking that they appeared on the scene only recently. In fact, General Motors introduced the first industrial robot, called Unimate, in 1961. Even now, about a half of the robots in industrial use are employed in automotive manufacturing, where the closely defined tasks and rigid environment perfectly match what robots are best at.[27]

Interestingly, while automobile plants routinely employ robots to install windscreens on vehicles, if a car owner suffers a damaged windscreen and goes to a car repair shop to have it fixed they will find that this work is done by a human technician.

In a different vein, despite huge amounts of money being expended on achieving this objective, it has so far proved impossible to develop a robot with sufficient manual dexterity to fold a towel. (Tying a shoelace is another example of something that is still beyond a robot's capacity.) Consequently, the much-vaunted employment of robots as independent domestic helpers, as distinct from tools to assist human helpers, is a very long way off, if, indeed, it is ever achieved.

Researchers in Singapore have been trying to teach an industrial robot to assemble an IKEA flatpack chair. The good news is that they have succeeded. The bad news is that it took two of them, pre-programed by humans, more than 20 minutes. It is alleged that a human could accomplish this task in a fraction of the time – although this particular human might well take much longer, or would give up, in frustration, before the task was finished.[28]

Google's research and development company/group, now known as X, recently ran a project to identify images of cats on YouTube. A *New York Times* article about this project ran under the headline: "How many Computers to Identify a cat? 16,000."[29]

In China, long hours and much treasure have been expended in trying to develop robot waiters that can not only take your orders correctly but can also be relied upon to bring the food to your table without pouring the soup into your lap. Three restaurants in Guangzhou in southern China that had made much of employing robots as waiters had to abandon them because the robots simply weren't good enough.[30] Doubtless, it has also proved difficult to get the robots to behave toward the customers with the same surly arrogance that their human counterparts display in many parts of the world.

Indeed, this is a central paradox of AI research. Tasks that seem very complex have proved comparatively easy for robots and AI to perform. But tasks that seem very easy have proved to be extremely difficult. This is often known as Moravec's paradox. In 1998 the roboticist Peter Moravec wrote: "It is comparatively easy to make computers exhibit adult level performance on intelligence tests or playing checkers, and difficult or impossible to give them the skills of a one-year-old when it comes to perception and mobility.[31]

This is closely related to what some refer to as Polanyi's paradox, after a remark made by the economist, philosopher, and chemist Karl Polanyi who observed in 1966: "We know more than we can tell." What he meant is that many of the things that humans do they do without following an explicit set of rules. The tasks that have proved most difficult to automate are those demanding judgment, common sense, and flexibility, in a way that humans cannot themselves explain. In that case, it is extremely difficult to codify such understanding in a way that AI can replicate it.

The not so impressive sides of AI

AI has great difficulty in dealing with instructions that are logically ambiguous, or even downright wrong. Take, for instance, the instruction seen in many lifts: "Do not use in case of fire." Most humans will readily understand that this means that *if* there is a fire you should not use the lift. But it can readily be interpreted to mean "do not use the lift at all *in case* there is a fire."

Or take the wonderful scene in the film *Paddington*, when the eponymous bear is about to take his first trip on the London Underground. He sees a sign on the escalator which will be familiar to millions of London commuters: "Dogs must be carried." Paddington quickly leaves the station in order to steal a dog so that he can carry one on the escalator, just as the instruction commanded. AI might well find that instruction as challenging as Paddington did.

AI enthusiasts will no doubt counter that it would be facile to program an AI to interpret such an instruction appropriately. I am sure that this is true, but it completely misses the point. Such logically ambiguous instructions may crop up without warning – and without the scope for preprograming – throughout life. Humans are able to deal with them precisely because they are not entirely logical.

Even the much-vaunted triumphs of Deep Blue have their limitations. For all his wonder about his conqueror, the defeated chess champion Gary Kasparov contrasted the earlier dreams of AI visionaries that they would create a computer that thought and played chess like a human, that is to say with creativity and intuition, with the reality that Deep Blue played chess like a machine and won through the exercise of brute number-crunching force, that enabled it to evaluate 200 million possible chess moves per second.[32]

The legendary MIT professor Noam Chomsky has put the achievement of AI in beating the reigning world chess champion into perspective. He said that this was no more surprising than a forklift truck winning a weightlifting competition. He might have added that many other creatures outperform humans in specific domains. For example, bats interpret sonar signals better than humans. But I have yet to meet anyone who thinks that, given sufficient time for evolution to play out, bats are going to surpass humans in general intelligence.

And the philosopher John Searle penned an opinion piece in the *Wall Street Journal* that wittily put the achievement of Watson in winning *Jeopardy!* into perspective. It appeared under the headline: "Watson Doesn't Know It Won on 'Jeopardy!'" Searle pointed out that Watson did not dream about it beforehand or celebrate it afterward. No chats with friends, no commiserations with vanquished opponents.[33]

Murray Shanahan has recognized the limitations. He has said: "A chatbot that is programed to crack a few jokes or a humanoid robot whose eyes can follow you around a room can easily give a contrary impression. But, as AI sceptics will quickly and rightly point out, this is just an illusion."[34]

For all the dramatic advances in AI, what has been achieved so far is just the electronic, digital equivalent of parrots or mynah birds which can often shock people with their delivery of words and phrases – but without understanding what they are "saying."

The limits to transformation

There appears to be a serious tendency among many AI experts to indulge in exuberant over-optimism. Professor Takahashi, one of Japan's leading robotics experts, warns that "people expect too much from robotics and AI." Referring to the dream of constructing an automaton that could be servant, companion, and general drudge, he says: "It's like putting colonies on Mars, possibly feasible technologically, but frankly not worth the massive investment it would take. There are better, more productive, ways of spending money."[35]

As to "The Internet of Things," rarely can something have been so overhyped. So, we will be able to monitor umpteen subjects in our everyday lives and tell whether they need renewing, polishing, cleaning, or mending. So what? I suppose there might be some helpful instances, but they will surely be peripheral, and will neither significantly reduce the demand for human labor nor meaningfully increase human wellbeing.

I recently had my first encounter with an "intelligent" toilet. Decency forbids me from reporting full details, but suffice it to say that the toilet's being able to speak to me, in a charming female voice, by the way, telling me its history and what it liked and disliked, added nothing to the utility – or enjoyment – of the experience.

In the future, doorknobs and curtains will also be able to speak to us when they need some attention, rather like those ghastly disembodied voices or noises in cars that tell us when we haven't fastened our seatbelts. Heaven forfend! I have a dystopian vision of all the objects in my life screaming at me in a cacophony of useless information. Such a vision is surely one of information overload in spades.

Such an avalanche of useless information will be far from costless. Back in 1971 the Nobel Prize winner Herbert Simon, whom we have already encountered, saw this coming. He wrote: "Information consumes the attention of its recipients. Hence a wealth of information creates a poverty of attention."[36] What wisdom! Poverty of attention is surely exactly what characterizes our age, dominated by the internet, the smartphone, and social media.

We have all witnessed tables in restaurants where the diners – friends, partners, lovers, or whatever – are glued to their smartphones and pay no attention whatever to each other. Being connected to anyone, no matter how distant, has distanced us from everyone, no matter how close. Heaven knows how much more extreme the lack of attention will be when all inanimate objects are able to add their "voices" to this cacophony.

Not the effects envisaged

Even where information technology has fulfilled the technological hopes entertained for it and has been employed in the workplace, it has still not had quite the effect on people and society that was envisaged – for both good and ill. There is a long history of people seeing the progress of technology as having negative economic consequences. In 1931 Einstein blamed the Great Depression on machines. In the late 1970s British Prime Minister James Callaghan commissioned a study from the civil service on the threat to jobs from automation.[37]

When they first emerged, it was widely predicted that computers would put an end to large numbers of office jobs. Nothing of the sort has happened, even though the job of typist has just about disappeared. And what about the paperless office? Remember that one?

In particular, it was widely believed when spreadsheet software appeared in the 1980s that this would cause huge job losses among accountants. By contrast, the number of accountants and auditors working in the USA has risen from 1.1 million in 1985 to 1.4 million in 2016. As so often before, the new technology has widened the scope of what accountants can do. Accordingly, there has been an increased demand for their services.

In technical economic terms, most popular accounts of the impact of AI follow the conventional wisdom about the impact of previous technological advances in assuming, rather lazily, that new machinery inevitably and always substitutes for labor. Some technological advances are indeed of this type, but many are *complementary* to labor and consequently increase the demand for it.[38]

There is another good example, much closer to home, of technological advance being wonderful and amazing yet less transformative than it at first seemed likely to be. Are you reading this in book form or on an electronic reader such as a Kindle? The majority of you, I am sure, will be reading it using the medium that has disparagingly been termed "dead trees." Yet only a few years ago it was widely predicted that e-readers would soon lead to the complete demise of the printed book. More than that, the end of printed editions and the transport of books to digital form would unleash the possibility of endless improvement. Consequently, a "book" would never be finished but would be constantly updated and revised. So, it would really cease to be a book. It would be more like a rolling news report.

What a nightmare! Pity the poor author. Imagine never being able to kiss goodbye to your baby and needing to revise and develop your "books" ad nauseam! I cannot think of anything worse.

But we authors can relax. This much mooted development has not happened. What's more, it looks as though it never will. Indeed, over the last couple of years, sales of e-books have been falling as readers have returned to the old printed version. I am not surprised. E-readers have their place but, on the whole, they do not match the convenience and the sense of the physical embodiment of ideas and thoughts that a printed book gives you. My guess is that the two will coexist but the printed version will retain the status of being "the real thing." Consequently, constant revision and the demise of the very idea of a book as a finished piece of work will live only in the realm of authors' nightmares.

Similarly, when cinema first appeared it was believed that it would before long kill off live theater. And later when television first appeared it was argued that it would kill off cinema. Nothing of the sort has occurred. Cinema, television, and live theater exist side by side. Indeed, they feed off one another.

The cost of robots and AI

What is at the root of the gloom and doom about future economic prospects spread by so many experts on AI, as well as quite a few nontechnological analysts, and even some economists? I think it is the vision of robots and AI as substitute workers. Once you see things in this way, you readily reach some gloomy conclusions.

Indeed, the whole robot and AI revolution begins to look like a rerun of what happened to the Western world thanks to globalization and the rise of China. These effectively added a few billion extra workers to the workforce but (initially at least) hardly any extra capital. The results were downward pressure on real wages in the West, lower prices, and a tendency to weak aggregate demand, leading to a regime of extremely low interest rates. This culminated in the Global Financial Crisis (GFC), followed by the Great Recession, the worst downturn since the Great Depression of the 1930s.

In fact, it is easy to understand why people thinking along these lines would believe that the robot shock will be worse than the China shock. After all, the opening up of China had such a major impact because Chinese workers were paid so little compared to their Western counterparts. Yet robots and AI "will work for nothing." No wonder people who look at things in this way are worried.

But this is not the way that we should think about robots and AI. They are pieces of capital equipment. As such, although they are not paid wages, do not receive benefits, and do not draw pensions, they are certainly not costless. They cost money to build, develop, and maintain, and they cost money to finance.

Moreover, the right software needs to be installed. That costs money. And the software needs to be, if not maintained, then at least updated, not only to keep up with latest performance norms but quite possibly even to be able to function at all. For the robot will need to communicate with other robots which will be constantly developing.

Admittedly, the Baxter robot works for about $4 an hour. But Baxters are, in fact, not very capable and are not in much demand. They may be cheap to run but they cost $22,000 and upward to buy. Sales of Baxters have not picked up, and in December 2013 Baxter's manufacturer, a firm called Rethink, laid off a quarter of its staff.[39]

According to Kevin Kelly, it costs $100,000 or more to buy an industrial robot but you may need to spend four times that amount over a lifespan to program, train, and maintain it, making a total bill over the robot's "lifetime" of half a million dollars or more.[40]

So, employing a robot will involve a fixed investment. And this investment will be subject to all the usual factors that govern whether an investment is worthwhile: the cost of the equipment and any maintenance costs, the rate of return, the cost of finance, and the risk, including the risk of obsolescence. This last factor could prove to be extremely significant. Even after robots have come to perform a major role in the economy, technological progress will continue. As both software and key design features of robots improve, older vintages of robot will lose value – perhaps even all value.

Think of the comparison with recorded music. You may have once been happy with the old 78rpm playing machines, but before long they became obsolete as 45rpm and 33rpm records took over. Subsequently, tapes and tape-playing machines had a brief moment in the sun before CDs took over. Now they are almost obsolete, along with the CD players that once seemed like the last word, as music is downloaded from the internet.

As long as robots cost something, then there is a chance for humans to outcompete them even if robots are more capable technically. And the higher the cost of robots, the easier it will be for humans to outcompete them. Or, to put this more technically, the higher the cost of robots, the higher will be the wage earned by humans that will still allow them to be the cheaper option.

Moreover, there is ample scope for the best competitor from a technical point of view to be a combination of humans and robots, or humans and AI. Take chess, for example. It has long been clear that AIs can outperform even the grandest of chess champions. But the evidence is now building up that a combination of chess champion and AI can outperform, not just the best human acting alone, but also the best AIs operating alone.

Wage flexibility

Whether humans, or robots, or a combination of the two, prove to be the cheapest and most effective option in a particular case will depend, like everything else, upon price. Certainly, the price of the robots and AI, including the capital and running costs, is part of this; but so also is the cost of human labor. The lower that cost is, other things equal, the more likely it is that humans will be employed rather than robots and AI.

During the recent Great Recession, because real wages fell, it seems that many British employers relied on the continued employment of people who could have been made redundant if they had spent money on upgraded computer equipment and software. For instance, law firms seem to have delayed investment in digital document management because legal assistants would do the work for very little pay.[41]

So, this means that how far and how rapidly robots and AI are employed is not just a technological given but rather responds to economic variables, in particular the level of wages and salaries, the rate of interest, and the cost of capital.

Mind you, this does not exactly provide a great deal of comfort. It seems to imply that, as robots and AI become more capable, humans will be able to retain their jobs only by working for less and less money. (The impact of the robot and AI revolution on the distribution of incomes is a subject that I take up in Chapter 6. And I discuss policies that might be deployed to ameliorate or offset any adverse effects on the distribution of income in Chapter 9.)

Looking at the issue this way, that is to say, by asking what the robotics and AI revolution implies for the demand for human labor, rather than thinking about the number of jobs for humans that will be lost, is the right way to proceed. This leaves the level of human employment (and unemployment) to be the outcome of how far prices (in this case, wages and salaries) adjust rather than volumes (i.e., the number of jobs).

But although this is the right way to think about the issue, it does not lead ineluctably to the conclusion that humans face a choice between less work and lower pay. Whether or not this happens depends largely on how robust the demand for human labor turns out to be. And, as I will show in later chapters, that depends a lot upon human tastes and preferences.

Comparative advantage

We can now bring all of this together with the help of a simple economic concept. At the root of exchange between both individuals and countries is what economists call *comparative advantage*. The essential idea is that even if one person (or country) is *absolutely* better and more efficient than another person (or country), at doing everything, it is better for both that the two should specialize in the activity at which they are *relatively* better and trade the surplus fruits of this activity for the surplus fruits of the other person (or country). Ever since the theory of comparative advantage was developed by David Ricardo in 1817, it has provided key insights into international trade. It has the hallmark of greatness: it is blissfully simple but utterly profound.

Some AI enthusiasts (and economic pessimists) argue that in the new world, as regards humans on the one hand and robots and AI on the other, comparative advantage will cease to apply. This is certainly the contention of the futurist and AI expert Martin Ford. He says:

"Machines, and particularly software applications, can be easily replicated. In many cases they can be cloned at a cost that is small compared with employing a person. When intelligence can be replicated, the concept of opportunity cost is upended. Jane can now perform brain surgery and cook simultaneously. So why does she need Tom at all?"[42]

Actually, this argument is not right at all. Unless and until robots and AI can produce and reproduce themselves costlessly (and that may not even be true beyond the Singularity), human beings will always have some *comparative* advantage, even if they are absolutely inferior to robots and AI in every task. (Mind you, that does not say much about the income that humans would earn in such a world. It might be appallingly low such that it hardly seemed worth working and the state has to intervene in a major way.)

But, in fact, such an outcome lies a long way off and, I suspect, will never transpire. For there are many areas where humans possess an *absolute* advantage over robots and AI, including manual dexterity, emotional intelligence, creativity, flexibility, and, most importantly, humanity. These qualities will ensure that in the AI economy there will be a plethora of jobs for humans. (I will sketch the future shape of the labor market in Chapter 5.)

Is it different this time?

So, we are now in a position to answer the question posed in the title of this chapter. Is it different this time? The answer is no. There are going to be some extraordinary developments and some dramatic improvements in the efficiency of doing customary things, as well as a whole host of new goods and services, many of which we cannot even imagine. And many people are going to lose, not just their jobs but also the very basis of their livelihoods. Moreover, robots and AI are going to reinvigorate the engine of growth which has seemed to sputter and stall over the last two decades.

So, the robotics and AI revolution will be momentous. But so were the steam engine, the jet engine, and the computer. And just as with other new technologies, there are going to be quite a few blind alleys, exaggerations, and disappointments as in some aspects robots and AI fail to deliver what they were hyped up to do, even while in others they exceed expectations.

This is reminiscent of the situation during the dot-com boom in the early 2000s. At that time it was thought that everything was going to migrate to the net; equally any madcap business idea was bound to make squillions for its originators if only the company name ended in .com. The madness of that time culminated in the bursting of the financial market bubble and the collapse of many businesses – as well as quite a few reputations. Yet the internet *has* changed the world. And some of the businesses that emerged in that frenzy of excitement and exaggeration have not only survived but have transformed the business landscape – Amazon and Google being preeminent examples.

We can already glimpse the shape of what is going to happen to the economy. There is a good chance that robots will continue to disappoint those who dreamed that they would be substitute workers, able to perform all manual tasks that humans can do without any human assistance or supervision. But we will probably be surprised by how much robots will be able to do in *assisting* human operators in both sophisticated tasks (such as surgery) and unsophisticated ones (such as social care). And most people still underestimate what AI can achieve in performing routine mental tasks. The effects are potentially devastating as a wide swathe of middle-class jobs are set to disappear.

Yet the changes taking place now, and due to take place in coming years, thanks to the increasing use of robots and AI, are in essence a continuation of the processes that have been taking place since the Industrial Revolution. And the effects will be broadly similar, too. First of all, and most importantly, the fourth industrial revolution, just like its predecessors, will greatly increase productive capacity. This will be unambiguously good. It will be up to us to decide how to make the most of the opportunities unleashed.

The parallels with *the* Industrial Revolution run wide and deep. Just as before, the Robot Age will not be positive for everyone. It may take a long while for everyone to be demonstrably better off. And, again like the major technological advances of the nineteenth and twentieth century, it may take a good deal of time for robotics and AI to have their full effect on the economy.

This is partly because, like many of their predecessors, the radicalism of the technological changes now afoot has been overhyped.

But it is also because, even where the changes are truly radical, they take time to be fully integrated with the rest of the economy and with society in general. Moreover, regulatory and ethical issues will result in the full implementation of new systems running behind their technical feasibility.

In particular, we must beware of thinking of the robotics and AI revolution as a sort of tsunami currently just beyond the horizon but advancing fast and set to engulf us some time soon. Robots have been in industrial use for more than 60 years. And early forms of AI have been in use for almost as long, contributing to improved productivity, and transforming employment patterns for years. Algorithms have been taking on jobs in things like customer service and bookkeeping. Such developments have been proceeding continuously and gradually yet, while radically changing some of the world of work, they have not produced the "transformation" that so many of the early enthusiasts told us was about to happen.

What is happening now is an intensification of these developments made possible by increased computer power, increased availability of data, and the development of the learning power of AI systems. The gradual and evolutionary nature of the changes afoot means that government, companies, and individuals have time – not time to do nothing, but time to consider and to make changes themselves in preparation for the changes in circumstances that will be thrust upon them.

Unanswered questions

This still leaves some extremely important issues to be addressed. They form the substance of the next part of this book. It tries to answer three key questions:

- To what extent will people take out the benefit of increased productive power in the form of increased leisure time rather than increased output?
- What sort of jobs will disappear, what sort of jobs will increase in number, and what jobs might spring up from nowhere?

- What will these changes do to the distribution of income between capital and labor, one sort of person and another, and one country and another?

Even when we have analyzed these major issues, our view of this terrain will still be only partial. For we have left out something vital – the macro aspect. In the world of robots and AI will there be the money to make the implicit demand for the services of human beings translate into effective demand that generates and sustains employment? And, even if there will be, can the transition to a new world take place without other disasters in the sphere of money?

After all, as I stressed in the last chapter, the history of the last 200 years has been marred by some major macro failures, of which the Great Depression of the 1930s was the most serious. More recently, we experienced the Global Financial Crisis, followed by what we have now come to call the Great Recession, which could have run the Great Depression pretty close. Much of the gloom and doom emanating from the AI geeks reflects a view about a likely malfunctioning of the macro-economy in a robot- and AI-dominated world, mirroring, but potentially surpassing, these previous disasters. Are they right?

To get the answer to that question, and several other related matters, we have to delve, albeit gingerly, into the world of macroeconomics. Readers may be forgiven for thinking that from what they have encountered of this subject before, this world is even more fictional than the musings of the AI geeks, and just as incomprehensible.

They have a point. Many modern economists have given economics a bad name. But the subject can be more literary and more comprehensible than AI geekery. Admittedly, I am biased, but I think that it can also often be more enlightening. Anyway, we have come to a stage where there is no avoiding it. In this sphere (and in many others) economics provides the essential pathway to understanding.

3
Employment, growth, and inflation

"Insofar as they are economic problems at all, the world's problems in this generation and the next are problems of scarcity, not of intolerable abundance. The bogeyman of automation consumes worrying capacity that should be saved for real problems."

Herbert Simon, 1966[1]

"Forecasting is extremely difficult – especially when it's about the future."

Mark Twain[2]

In his trademark lofty tone, the legendary economist John Kenneth Galbraith once pronounced: "The only function of economic forecasting is to make astrology look respectable."[3] As an economic forecaster myself, even if I think Galbraith's view is an exaggeration, I have considerable sympathy with it.

In that case, what is to be said about forecasting the *distant* economic future? It is extremely hard to see ahead with any degree of confidence as to direction, never mind anything remotely amounting to accuracy as to magnitudes, what long-term trends in the economy will be. I tried to do this with *The Death of Inflation*, published in 1996, in which I envisaged an extended period of low inflation, dipping sometimes into deflation, accompanied by ultra-low interest rates, for virtually all countries of the developed West.

If you will forgive me, it is tempting to say that subsequent years have broadly vindicated this prognosis. But I managed to make such a radical forecast because a few powerful forces were pointing in the same direction at once and in the ensuing quarter-century no major countervailing force appeared. It could easily have gone the other way.

With regard to the economic effects of robots and AI, things are not so straightforward. There are many possible countervailing forces. Moreover, the public policy response to the stresses and strains that will be created by robots and AI remains unclear. Accordingly, we have to grapple with profound uncertainties. Consequently, what I have to say must be hedged about with qualifications and caveats.

This is particularly true since, as I argued in the last chapter, both the magnitude of the economic impact of robots and AI and the length of time over which this impact will play out remain keenly disputed. In what follows I concentrate on establishing the *direction* of the likely macroeconomic effects, not least because this can sometimes be surprising. And this analysis can be helpful whether the scale of the effects is enormous, as some people assert, or nugatory, as others suspect. And it can be helpful whether the impact is going to be immediate, sudden, and intense, as many experts on AI believe, or drawn out over several decades, just like the impact of the previous major advances from the steam engine to the computer, as others believe.

But wherever possible I make my own attempt to attach some likely magnitudes to these effects, and to put them into the context of the historical experience, which I reviewed in Chapter 1.

The macroeconomic effects

There are eight main areas where the spread of robots and AI may have a decided impact on the macroeconomy:

- The overall level of economic activity and employment.
- The rate of inflation.
- The pace of economic growth.
- The level of interest rates.
- The performance of different assets.
- The balance between work and leisure time.
- The type of jobs that will be available.
- The distribution of income.

Because they are both large in scope and relatively discrete, I have decided to deal with the last three topics in three individual chapters after this one. But in what follows here, I will discuss each of the first five in turn. I should say at the outset, however, that all these aspects of the macroeconomy are interrelated – including the three topics that I am holding over to later chapters. Everything feeds into everything else.

As I warned in the Prologue, this leads to an acute problem of both understanding and presentation. Even though, in practice, all things are determined together, with all interrelationships playing out simultaneously, this is not how we can analyze the issues at stake. If you tried to think in this way, you would send yourself round and round in circles.

What I do here, in time-honored fashion, is to break things into parts and to concentrate on one part at a time, leaving other questions, and their interrelationship with the issues at hand, to be examined later. I do my best to refer to these interrelationships as we go along, but everything isn't fully drawn together until the Conclusion. I start by sketching out a general macro perspective on the robot and AI revolution and then I look at the prospects for the overall level of aggregate demand, and hence employment and unemployment, before going on to look at, in turn, the other four topics that I mentioned above.

More productive capital

In the last chapter I made it clear that robots and AI should be regarded, and analyzed, as types of capital investment. Because of some remarkable technological developments, robots and AI are becoming much more capable and hence productive. Investment in these things has become more rewarding and, if the enthusiasts are right, will become still more rewarding. (Economists will refer to this as an increase in the marginal efficiency of capital.)

Once you look at things in this way, traditional economic analysis can be brought to bear. There are several clear implications. The increase in the return on capital should lead to:

- An increase in investment.[4]
- Upward pressure on real interest rates.
- An increase in real output and income per capita.
- A possible increase in average real wages.

This last result is possible because, since more capital has been added, the amount of capital per worker will have risen. Yet whether this leads to an increase in average real wages will depend, just as with any other capital investment, on the extent to which, at the macro level, the new capital substitutes for labor as opposed to complementing it. So much of the AI gloom-and-doom literature implicitly or explicitly assumes that robots and AI are pure substitutes for human labor. Yet, as the discussion in the previous chapter indicated, and later chapters will confirm, in many parts of the economy robots and AI are complementary to human labor.

Human and artificial intelligence are fundamentally different and they excel at different things. What's more, this is likely to continue. So, humans and robots and AI will produce more by working in cooperation with each other. But, since it is a human world, and robots and AI have no independent existence, or claim to reward (at least until the Epilogue), it is humans that will benefit.

So, armed with this perspective, we are now in a position to think through the various macroeconomic effects of the robot and AI revolution, starting with the impact on economic activity and employment.

Economic activity and employment

The techie literature is full of visions in which technological progress leads to economic Armageddon, and specifically to a future of mass unemployment and poverty. It is not easy to unpick the arguments being advanced. But we must try.

In a modern economy, there are several different types of unemployment, including *frictional* unemployment caused by people moving between jobs not being able (or willing) exactly to coordinate leaving the old job with starting the new one, and *structural* unemployment, where the industries and/or areas in which people work go into decline, perhaps involving even the redundancy of the very skills on which they built their livelihoods. Although the robot and AI revolution will involve elements of both these types of unemployment, on their own these factors are not large enough to lead to the apocalyptic vision that many AI luminaries have described.

There are two versions of their ultra-pessimistic vision. They are interrelated, but they are nevertheless distinct. The first is in essence technological. It is the view that there will be next to no jobs that humans can do better than machines. Accordingly, there will be few jobs performed by humans and mass unemployment will be inevitable.

The second is in essence economic. It is the view that the spread of robots and the development of AI will rob the economy of purchasing power. So, even if, technologically speaking, there could be jobs for humans to do, there would not be the demand in the system to enable them to be employed.

I have already addressed the first version of this vision in the previous chapter, and found it wanting. I indicated how and why mass employment can continue. And in Chapter 5 I will lay out a vision of what the jobs of the future might be. But I must tackle the second, economic, version of the pessimistic vision here.

This negative vision is nothing new. On a visit to a Ford manufacturing plant in the 1950s, the American Union boss Walter Reuther saw an impressive array of robots assembling cars. The executive who was showing Reuther round asked him how he thought the robots would be made to pay union membership fees. Reuther replied that the bigger question was how the robots would be made to buy cars.[5]

With regard to the economic impact of robots and AI, there are many uncertainties about the future and many risks. Is the prospect that there will be no demand to buy the output of the robots one of them? It is time for the basics of Economics 101. I will first lay out briefly the simple economics of aggregate demand without reference to robots and AI. Then I will bring them in, analyzing how they fit into the theoretical framework just established.

So, with hot towels at the ready, here goes. The place to start is with a simple statement: supply creates its own demand. This is to say, if output is produced, someone somewhere has the income to spend on purchasing it. This was the dictum of the great French economist Jean-Baptiste Say, who was writing in the early nineteenth century.[6] And Say's Law, as it is known, is as true today as it was then. I have chosen my words advisedly, because Say's Law wasn't absolutely true even then – and it isn't absolutely true now. So, it isn't only gushing tech enthusiasts who overuse the word "law" to describe a relationship. Economists are sometimes guilty of this, too.

Still, let's start with the simple truth of what Say's Law is about before dealing with the complication. For the macroeconomy, production equals income which equals spending. These are, if you like, three sides of the same coin. If things are produced, then things are there to be bought by people who have the income to buy them as a result of being paid to produce them. This is still true in a robot- and AI-driven world. The robots may not have spending power but whoever owns them does. So, if robots and AI increase productive capacity, then they increase incomes and consumption capacity – for someone. As to who that someone might be, I'll discuss that in a moment.

Keynesian conditions

But now let us consider the complication. In a money economy, although production gives rise to income that can be spent to buy what has been produced, in fact this income may not be fully spent. And if it isn't, then some of the things produced will go unbought, which will lead to production being cut back and people being laid off, which will reduce income, which will reduce the capacity to buy things, and so on and so forth. This is the description of an economic downturn. Recovery happens when this process goes into reverse. Economic downturns do not *need* to happen because the income is there to be spent. But they *can* happen because income will not necessarily always be fully spent.

In practice, in a money economy there are always fluctuations of this sort.[7] In normal conditions, such ups and downs are minor and temporary. In really serious economic conditions, however, demand can stay depressed for a considerable period of time. John Maynard Keynes explained how this could happen, and he laid out what could be done to overcome such depressions if and when they occurred. In these conditions, Keynes advocated government and central bank action to restore aggregate demand to normal.

Until the depression in Japan in the 1990s, and more recently the Global Financial Crisis of 2007/9 and the subsequent Great Recession that enveloped most of the developed world, these depressionary conditions and Keynes's advice about how to deal with them were widely regarded as quaint, a sort of historical curiosum of concern only for those interested in economic history, and particularly in the Great Depression of the 1930s.

But after the events of the last 10 years not many economists would take this view. Keynes is back. Although there continue to be marked disagreements among economists about policy details, it is now pretty much the accepted wisdom among policymakers and academics that not only is there the possibility of governments and central banks taking action to prevent and, if necessary, to correct pronounced shortfalls of aggregate demand, but also it is their duty to take such action.[8]

The candidate measures include increases in government spending, reductions in taxes, cuts in interest rates or increases in the money supply through the policy that has become known as quantitative easing (QE). In the end, if all else fails, there is the option of distributing money to the people gratis – the so-called "helicopter money," first referred to by Professor Milton Friedman, and recently discussed and advocated by, among others, the former chairman of the Federal Reserve Ben Bernanke.

This does not mean that recessions – or even mini-depressions – cannot occur. But it does mean that a really serious depression, such as occurred in the 1930s, is unlikely – unless the authorities take leave of their senses and for political, or ideological, reasons fail to act with sufficient vigor.

Depressive tendencies

The above conclusion about macro policy action applies just as much in a robot- and AI-dominated future as under any other conditions. But will the need for such action be any greater in the Robot Age? That is to say, in the new world will there be a greater tendency toward underspending and therefore high unemployment, which the policy authorities have to counter in the various ways mentioned above?

In this question I do not refer to the idea that there won't be many jobs for humans because they will all have been taken by robots. As I hope I made clear in the previous chapter, as long as human beings still want services, and to some extent goods, provided by other human beings, there will be demand for human labor. (I will discuss the extent to which they will want these things, and what the jobs of the future might be, in Chapter 5.)

No, the tendency toward unemployment that I am going to discuss here is, if you like, Keynesian unemployment. This could potentially affect robots as well as humans. This possibility arises as a result of the implications of robots and AI for the level of aggregate demand.

There are two plausible reasons why the AI economy may tend toward depressed aggregate demand. The first is that, unless something specific is done to redress this tendency, in the new world a larger share of the national income may tend to accrue to the owners of capital, including robots. In other words, there will be a shift in the distribution of income away from wages and toward profits. Unless companies have the same propensity to spend profits, and/or that when they are distributed to their shareholders, *they* have the same propensity to spend their dividends, as wage-earning consumers do to spend their pay, then lower aggregate demand will result.

The second possible route to this same result comes from wage earners themselves. Whether or not robots and AI lead to a shift in incomes toward profits, they may increase the income gap between workers (and potential workers) who have few skills, or skills that are easily replaced by machines, and those workers who possess valuable skills, in particular those who can readily work alongside robots and AI and hence enjoy increased productive capacity. Indeed, the wage that the worst-placed humans can earn in this machine-dominated labor market could be so low that large numbers of people choose not to work since their potential income from employment falls below what they can receive from the state in benefits.

In other words, the robot and AI revolution could make the distribution of income more unequal. Whether or not this shift is deemed to be desirable, socially acceptable, and politically sustainable is one set of issues. (I discuss these matters in Chapter 9). But, over and above this, because people with lower incomes tend to spend a higher proportion of their incomes than richer people, this more unequal income distribution, if it happens, could result in a tendency for aggregate demand to fall short of productive potential.

Is deficient demand likely?

On the face of it, it does seem plausible that the increased employment of robots and AI will lead to increased inequality of incomes between workers. (I discuss this issue in Chapter 6.) Equally, in the first instance, without any deliberate policy action by government to spread the benefits accruing from the employment of robots and AI (perhaps through the imposition of a robot tax whose revenues are used to fund a universal basic income, which I discuss in Chapters 7 and 9), the impact will probably also be to boost profits at the expense of wages.

But even if one of these two things does happen, or even both, we cannot blithely assume that they will inevitably lead to deficient demand. For a start, if the robot and AI revolution is as profound as its enthusiasts allege, then society as a whole will be radically changed. It would be unwise to launch into a confident assertion of the impact of income distribution on aggregate demand based upon experience in a completely different state and structure of society. This is exactly what happened after the Second World War when the economics profession blithely assumed that the end of war production would lead to a return to the demand deficiency of the 1930s. In the event, an upsurge of investment and consumer spending led to strong aggregate demand.

Moreover, in the Robot Age, whatever happens to income distribution, there are likely to be some significant offsetting factors on aggregate demand. Over the last 20 years, the world has experienced two powerful forces that have weakened aggregate demand. First, because populations have been aging, the demographics in much of the West have favored saving over consumption. Very soon, however, in many Western countries the age balance of the population will have changed so much that there will be a large increase in the number of retirees relative to workers. Not only will this reduce labor supply, but retirees typically spend a high proportion of their income and indeed tend to run down their accumulated savings. Accordingly, in many Western countries, demographic change will soon swing toward reduced saving and increased spending, and hence stronger aggregate demand.

Second, in the years running up to the GFC and continuing for a while afterward, there was a significant imbalance in the world between high-spending (and low-saving) countries, such as the USA and the UK, and low-spending (and high-saving) countries such as China and the oil producers. The former tended to run large trade deficits whereas the latter tended to run large trade surpluses. As always, the deficit countries were put under pressure to reduce their deficits by reducing spending whereas there was no corresponding pressure in the opposite direction in the surplus countries. If you like, at the global level there was a tendency toward deficient spending (excessive saving) which exerted deflationary pressure on the world economy. This was countered by expansionary monetary policies that produced distinct problems of their own, including contributing to the factors that caused the GFC.

But, recently, one source of these international imbalances has faded to the point of insignificance, namely the surpluses accruing to the oil-producing countries. As the price of oil has fallen dramatically, so the surpluses of the oil-producing countries have fallen almost to zero. (Mind you this can always change and, indeed, may already have done so by the time that you read this.) Moreover, the Chinese surplus has also fallen significantly.

The reduction in global trade imbalances should help the stability of the world economy and help the maintenance of aggregate demand without resort to unsustainable monetary policies. The major continuing source of international imbalances is now the eurozone, and especially Germany. But, one way or another, I suspect that this is going to change as well. (An analysis of the eurozone and its contributions to global imbalance is beyond the scope of this book. Readers interested in this question could consult my book *The Trouble with Europe*, recently revised and republished as *Making a Success of Brexit*.[9])

Meanwhile, the banks are gradually recovering from the near-collapse of 2007/9 and the subsequent onslaught of regulatory restrictions. They are becoming more able and willing to lend. This, too, supports aggregate demand.

Over and above the fading of the factors that have weakened aggregate demand, I could readily see the emergence of a major new force that will strengthen it. The new opportunities unleashed by the spread of robots and AI, along with advances in nanotechnology and biotechnology, should provide a wide array of investment opportunities. I could readily imagine that the AI economy could be characterized by buoyant investment spending. Accordingly, even if the robot and AI revolution does shift income away from wages and toward profits, far from the vision of endemically deficient demand, this need not necessarily lead to weak aggregate demand. Perhaps quite the reverse. (I discuss this in more detail below.)

The policy response

But there is also macroeconomic policy to contend with. Suffice it to say that if, despite the offsetting factors described above, the age of robots and AI has a tendency toward depressed aggregate demand, matters would not rest there. The policy authorities would intervene to boost aggregate demand in some or all of the ways mentioned above.

Admittedly, there would probably be a particularly tricky transition problem. If, in the initial phase of development, large numbers of human beings found themselves unemployed, then they would not have the incomes to be able to spend on new services provided by other human beings. But this too can be eased by public policy. It is a transitional problem and not the final destination.

Accordingly, even if the above argument about the likely effects on the distribution of income and hence on aggregate demand is correct, and there are no offsetting factors, then it is wrong to conclude from this that this is bound to be an epoch of deficient demand and high unemployment. Rather, in these circumstances the onset of the Robot Age would be likely to be the age of expansionary policy. This could include expansionary fiscal policy, QE, or helicopter money. But the first resort of policymakers would surely be the imposition of ultra-low, or even negative, interest rates. (I will discuss likely future interest rates and bond yields in a moment.)

Mind you, there are limits to this cocktail of expansionary policies. Sustained budget deficits produce an accumulation of public debt that potentially poses severe problems, including the financial solvency of the state. If debt rises above sustainable levels, the eventual outcome is likely to be either default or inflation or both. Accordingly, although an expansionary fiscal policy may be a viable way of addressing deficient demand for a temporary period, it is not sustainable in the long run. So, if deficient demand is the normal state of affairs in the AI economy, expansionary fiscal policy does not offer a way out.

Ultra-low interest rates would probably be more sustainable but they carry a series of risks and costs. Most importantly, ultra-low rates sustained for a long period would severely distort both financial markets and the real economy. This would potentially impair the economy's productive potential. Moreover, by boosting asset prices, sustained low interest rates tend to lead to a more unequal distribution of wealth. Similar points apply to a sustained policy of QE.

The rate of inflation

I frequently come across people who believe that the age of robots and AI is bound to be an epoch of low inflation, or even deflation (i.e., falling prices). I wish I had their confidence. In practice, this view is much too simplistic. Admittedly, if a shift in the distribution of income delivers a tendency toward aggregate demand falling short of aggregate supply, and the potential offsetting factors, discussed above, are not powerful enough, the consequence would be a tendency toward unemployment of both human labor and capital, including robots. This would lead to downward pressure on prices, which would result in lower inflation than there would otherwise have been. Accordingly, you can easily see how people might conclude that the Robot Age will also be the age of low inflation, or even deflation.

But there are three big "ifs" here:

- Whether the robot and AI revolution does have a marked effect on the distribution of income.
- If it does, whether this leads to a tendency toward deficient demand.
- If this happens, whether offsetting factors, including the policy response, are inadequate.

What should be acknowledged, though, is that the initial effect of the introduction of robots and AI is to reduce prices. After all, robots and AI are introduced to reduce costs, predominantly by saving human labor which is released into the labor force, or not taken up in the first place.

This mirrors the effect on the world economy from the advent of globalization and the emergence of China as a large producer of cheap manufactured goods. This imposed a downward shock – or rather a series of rolling downward shocks – to the price level, and thereby helped to keep the inflation rate down.

But this would be – as it was with globalization and the rise of China – essentially a transitional effect – albeit a pretty extended transition. Moreover, as I argued in *The Death of Inflation*, in the end the rate of inflation will be the outcome of economic policy, and especially monetary policy. If the authorities do not want a regime of constantly falling prices, or even a regime of very low inflation, and they are prepared to take tough enough action to prevent this, then we will not have a regime of falling prices or very low inflation. This applies as much to any price deflationary tendencies unleashed by the employment of robots and AI as it did to the effects of globalization and the rise of China.

There is no reason to suppose that the monetary authorities, and behind them the state of popular opinion, expressed at the ballot box, will opt for high inflation, low inflation, or any particular rate of inflation any more in the Robot Age than at any other time.

Inflation and unemployment

That said, just as with globalization and the rise of China, the AI revolution could influence the way that inflation behaves, and specifically the relation between inflation and unemployment. Over and above the direct impact on the price level from globalization and the rise of China, over the last 25 years, the digital revolution has brought about a whole new economic reality at the micro level that theoretical economics has been struggling to understand and to incorporate into its models of the way the economy works.

From time immemorial, shortage has been the abiding feature of our lives – shortage of food and shelter, shortage of land, shortage of tools, shortage of everything. Indeed, the whole subject of economics grew up as a response to shortage, as a way of working out how to manage and cope with the various shortages in order to get the best result. And the conceptual language of economics is dominated by the continued pressure of shortage. Take, for instance, the concept of "opportunity cost." The opportunity cost of doing something – anything – is the forgoing of whatever else you could have done with the money/time/attention/involvement or whatever it is that you have devoted to the thing, service, or use of time that you have chosen.

Yet the IT/digital revolution massively expanded the range of things (or more usually non-things) that do not obey the usual laws of shortage. Anything that can be digitized, such as information, data, or knowledge – is not diminished by being shared more widely. The term used to describe this quality is "non-rival." Moreover, in the digital world network effects abound. The marginal cost of increasing the membership of a network is virtually zero but it has benefits for all other members of the network.

Economics has grown up to analyze a world where goods are made up of atoms. But things made of bits display four markedly different characteristics:

- They can be replicated perfectly.
- Extra units can be "produced" at (virtually) zero marginal cost.

- They can be transmitted instantly.
- They can be transmitted at (virtually) zero marginal cost.

The AI revolution will extend and intensify the aspects of the economy that display these characteristics. There will still be shortages, of course, of land and, linked to this, of position, as well as of food and other material goods. And humans will still be time-constrained. But the size and scope of the digital, networked, abundant world is set to increase relative to the material, limited, shortage-defined world.

This will influence the overall behavior of costs. They will continue to rise as output expands in the material, shortage-defined world. But in the digital, networked world, as output rises, average costs fall; and marginal costs do not rise and may even fall.

The interaction between these two parts of the economy will influence relative income shares, profit rates, and the behavior of the macroeconomy. A potentially negative consequence is that falling cost curves tend to favor the growth of monopoly and/or oligopoly and to reinforce "first-mover advantage." This influence is most evident in the information technology industry itself. The internet has given birth to enormously profitable and influential corporations with startlingly small workforces. (I take up these issues in Chapter 6.)

But a more favorable implication is the continuation of stable inflation at higher rates of economic activity and lower rates of unemployment. (If you have a taste for jargon, in Economese this is referred to as a flattening of the Phillips Curve, which is the curve depicting the relationship between inflation and unemployment.)

The pace of economic growth

I argued in Chapter 1 that the notion that there has recently been a profound and long-lasting slowdown in productivity growth was unconvincing. Quite apart from whatever opportunities are thrown up by robotics and AI, I fully expect the rate of increase of productivity growth to pick up from the low levels that have recently been registered in the developed world. That is, of course, provided that productivity growth is measured properly, which is a big "if."

So, the backdrop to considering the likely economic growth rate in the decades ahead is of productivity growth in the developed countries of perhaps 1–2 percent. Now let us consider what impact robots and AI might have on this already-improving picture.

In the first three decades after the Second World War, in much of the Western world, productivity increased at an average annual rate of over 3 percent. There were several reasons for this good performance but the scope for capitalizing on the technological advances that had built up over the previous 20 years, and the high level of investment made possible by the combination of full employment and low interest rates, were prime among them.

Something similar may be about to happen again. There will be two main factors driving such a surge in productivity growth. The first is the direct replacement of humans by "machines," including AI, mirroring what has gone on now umpteen times during the last 200 years. This will occur across a broad range of activities.

The second comes from the use of robots and AI to help human workers in a wide array of service activities. These will enable the human service operator to get more done in a given amount of time. This will apply just as much to care workers now employing robots to increase what they can achieve during the time allocated for a visit to the home of an elderly person, as to surgeons, now operating with the aid of high-level robotics. (In this latter case, the use of robots will not necessarily do much to speed up the surgery, but it will make it more accurate and less invasive, as well as allowing the surgeon to operate remotely, thereby potentially saving resources of time spent on travel.) I will review the scope for these effects in Chapter 5.

The potential productivity gains in the service sector are more significant than they may seem. For decades now, productivity growth has been much faster in manufacturing than in the service sector. As economies have become increasingly dominated by services, this has been a leading factor restraining the growth of productivity, and hence living standards. Indeed, the appreciation of this factor's importance has underpinned a good deal of the pessimism about the potential for productivity growth in the advanced economies.

The essential feature of service industries that has restricted their capacity to generate productivity growth has been the limited ability to augment the productive power of labor with capital equipment. Whether the service sector was education, medical, or social work, this essential feature held sway. But now the spread of robots and AI offers the scope for a significant increase in the amount of capital deployed in the service sectors and hence for substantial increases in productivity.

Jim Al-Khalili, Professor of Physics at Surrey University and President of the British Science Association, has recently come up with the estimate that by 2030 AI could add $15 trillion to global output, which is more than the combined current output of China and India.[10]

I am not sure where he gets this estimate from, but a little arithmetic will put his claim into perspective and help us to gauge how significant robots and AI could be. Let us suppose that the AI revolution drives up the average annual growth of per capita GDP (a proxy for productivity growth, at least over the medium to long run) to what was achieved in the period 2000–7, immediately before the GFC. For the developed countries that rate was 1.6 percent per annum and for the world as a whole it was 3.1 percent. If these growth rates were sustained for 10 years, then the cumulative increase in GDP per capita would be just over 17 percent and 35 percent respectively. But if these rates of growth were sustained over 30 years then the cumulative increases for per capita GDP would be 61 percent and 150 percent respectively.

The growth rates were higher, of course, during the Golden Age between 1950 and 1973 (and the ranking between the developed economies and the world as a whole was reversed). If we assume a return to the growth rates registered during the Golden Age (1950-1973) by the world as a whole, this would give a total increase in per capita GDP over 10 years of 32 percent, whereas 10 years at the rates recorded by the developed countries would give cumulative growth of 37 percent. The 30-year growth figures are pretty spectacular: nearly 130 percent and over 157 percent, respectively.

Combined with other forces at work, including nanotechnology and the recovery from the GFC, something like this rate of expansion should be perfectly feasible. Imagine GDP per capita, and hence living standards, more than doubling in a generation. This really would be a spectacular result and would fully justify the enthusiasm of the AI junkies while confounding the pessimism of the AI gloomsters.

From productivity to living standards

Yet simply saying that productivity growth may be rather high in the Robot Age does not translate directly into saying that economic growth will occur at the same rate, nor indeed that it will be high by recent standards. There are two bridges that have to be crossed before we can make this jump. First, there is the point about aggregate demand that I made above. It is possible that, although productivity growth is high, increasing numbers of people (and robots) will remain unemployed, or underemployed, as aggregate demand fails to take up the now rapidly increasing supply capacity. As I argued above, however, I think this prospect is unlikely, not least because there are policy measures that can be taken to prevent it from being realized.

The second factor is both larger in significance and more likely to be realized, namely the tendency for increased supply capacity to be taken out in the form of increased leisure rather than increased output. In the developed countries at least, people won't want double the quantity of the things that they currently consume. There will doubtless be new things that will be the object of their desires. Or perhaps they will choose more leisure over more things. That opens up a whole new set of questions that I will examine in the next chapter.

To the extent that this happens, the measured rate of increase of GDP, which will be below the measured rate of increase of productivity, will not do justice to the improvement in the human condition. It is perfectly possible that people will come to prefer increased leisure so much that, despite a significant increase in productivity, measured rates of increase of GDP are unspectacular. That said, on past form, as I show in the next chapter, the most likely outcome is a combination of increased output and increased leisure.

Let us be clear. If anything like the vision of more rapid growth of productivity expounded here is realized, then the result will be a faster rate of economic growth than has been normal in the West over recent years, and a faster increase in living standards, even on the inadequate measurements that we have, such as GDP or real incomes, or real consumption per capita. Moreover, because of increased leisure time, the underlying rate of improvement in the human condition will be still greater.

Interest rates

What does this vision of the future imply about interest rates? It is usual for economists to distinguish between nominal (i.e., money) rates, that is to say, the rates that we pay on loans or receive on deposits, and so-called *real* rates, that is to say nominal interest rates minus the rate of inflation. They make this distinction because, as the nomenclature indicates, it is *real* interest rates that should influence economic behavior. Indeed, if you look exclusively at nominal interest rates you can easily be led up the garden path. (This is not to say, though, that nominal rates are without importance altogether.) Accordingly, I am going to uphold this distinction here. I begin by discussing real interest rates and then go on to say something about nominal rates.

Amid all the sound and fury that swirl about concerning what determines real interest rates, usually signifying nothing, there are only two major factors that have a significant bearing:

- Whether aggregate demand falls short of, or exceeds, potential productive capacity.
- The rate of return on productive investment relative to the supply of savings to finance investment.

The first factor determines whether the monetary authorities are persuaded to try to boost demand by the imposition of low real interest rates, or to restrain it by the imposition of high real interest rates. Naturally, their writ only runs over the short-term interest rates that are directly under their control. (In the UK, this is Bank Rate; in the USA it is the Federal Funds Rate.) But, in practice, these short-term rates tend to exert a major influence on longer-term rates of interest, and hence on yields and required rates of return across the whole investment spectrum.

The second factor is quintessentially long term. It corresponds to the factors that the classical economists thought determined "the" rate of interest. It is a balance of forces, and the outcome for real interest rates will depend upon both elements, that is to say, both the propensity to save and the inclination to invest.

I have already in this chapter discussed the influences on the policymakers setting interest rates. Among those who have thought about the issue, and implicitly for almost every AI specialist, there is a presumption that the Robot Age will be characterized by underemployment and depression, as discussed above. Accordingly, whether they realize it or not, given that policymakers would try to counteract this, their vision of depression and unemployment in fact amounts to a vision of sustained low interest rates – at least until worries about the distortionary consequences of such a policy persuaded the authorities to abandon it.

Their vision could be right. But whenever I see a conventional wisdom lazily established yet deeply entrenched, everything in my experience urges me to look in the opposite direction. As I argued above, even if public policy does not forestall, or at least limit, the possible shift in the distribution of income against wage earners and consumers, toward profits, and away from low earners toward high earners, aggregate demand could plausibly turn out to be strong rather than weak. I suspect that it will be.

If I am right about this, and if the combined effects of aging, a better-balanced world economy, stronger banks, and a surge of AI-related investment is large enough to offset whatever weakness of consumption arises from increased inequality, then there would be no reason for the Robot Age to coincide with a period of ultra-low interest rates. Indeed, there is a good case for believing that, under such conditions, short-term real interest rates should return to somewhere near the level that was believed to be normal before the onset of the GFC in 2007/9 – or perhaps even higher.

Savings and the return on capital

It is now time to say something about the second set of factors that determine real interest rates, namely the balance between savings and the demand for capital. It is possible that, for the reasons given above and developed in later chapters, the new world will be one given to a high rate of savings. Other things equal, that would depress real rates of interest. But this scenario is far from guaranteed. It would be unwise to take high rates of saving as a given. In particular, as mentioned above, demographic changes, involving a substantial aging of the population in Japan, China, and Europe, as mentioned above, may push in exactly the opposite direction.

Meanwhile, as argued above, the robotics and AI revolution potentially equates to a huge increase in the return on capital, and hence in the demand for long-term funds. Other things equal, this will push up real interest rates.

On balance, I suspect that the fantastic opportunities unleashed by robots and AI will lead to a surge in investment and this will both serve to keep aggregate demand strong, hence undermining the case for low real short rates in order to boost demand, and to absorb the pool of available savings. There is no certainty about this, but my bet is that the Robot Age will be an age of comparatively high real interest rates, not only at the (policy-determined) short end, but across the whole gamut of interest rates.

Nominal interest rates

Most people think, and deal, not in real interest rates but in nominal – that is, money – rates. What is to be said about them? The difference between the two is, as mentioned above, the rate of inflation. (Strictly speaking, it should be the *expected* rate of inflation.) As discussed above, I see no reason to believe that the AI economy will be characterized by high or low inflation. How things turn out with regard to the prevailing rate of inflation will largely lie in the lap of the monetary authorities and, behind them, the political forces playing out in society and expressed at the ballot box. Accordingly, what I have said about the likely course of real interest rates can be taken, as a first approximation, to apply to the likely level of nominal (i.e., money) interest rates as well.

This does not amount to a prediction of low inflation – and hence low nominal interest rates – in the AI economy. Rather, once the transition phase, which will be prone to downward shocks to the price level, is largely over, it is the assertion of an agnostic position. Inflation may turn out to be high or low in the AI economy. All I am saying is that, once the transition phase is over, I can see nothing in the nature of the changes brought about by robots and AI that can make the economy systematically inclined to one result or the other.

But a word of warning is in order. The range of possible variation in real rates is pretty small, perhaps something like 6 percent or 7 percent – from about minus 1 percent or 2 percent to about plus 5 percent. And even that range is very high in the wider scheme of things. In practice, real rates usually lie somewhere in the range between 0 and plus 3 percent. By contrast, the range of possible inflation rates is unlimited. In practice, inflation has been minus 20 percent to plus several thousands of percent. Accordingly, what I have said about real interest rates could easily be swamped in its effects on nominal rates by changes in the inflation environment.

The performance of different assets

What happens to interest rates is going to be one of the most important influences on the behavior of different assets and hence on investment returns in the Robot Age. What I have to say here cannot, and is not intended to, be any sort of investment tip sheet. How could it be?

But readers have a legitimate interest in how different types of assets can be expected to perform on average, and in aggregate. Moreover, their performance may even have some macroeconomic impact. So, it is important to consider what may happen to different types of asset.

If, for the reasons given above, real interest rates turn out to be low, then this will have an overwhelming influence across more or less all investment categories, supporting asset values but also continuing the era of low rates of return. But, again for the reasons given above, my suspicion is that this will not transpire and that real interest rates will return to something like the level that was regarded as normal before 2007–9. In that case, what can be said about the behavior of different assets?

What happens to bonds will be closely governed by the level of real interest rates, both in the present and the level expected in the future, as well as by the expected rate of inflation. As indicated above, I see no reason to suppose that the widespread use of robots and AI will push the average inflation rate in either direction, and I have no reason to believe that any such result would be expected by the financial markets. I have already said that a return to something like "normal" real interest rates, or perhaps higher, is to be expected. In other words, a return to something like normal bond yields, or perhaps higher, is to be expected also. From the current position that implies significant capital losses for existing holders.

Meanwhile, equities are likely to be buffeted by opposing forces. On the one hand, there will be a depressive influence originating from what is likely to happen to bonds, as outlined above. On the other hand, though, if I am right in supposing that economic growth will be strong, then this will have the effect of boosting the rate of growth of profits. Moreover, even without such a boost, there is the real possibility, described above, of a substantial shift in the distribution of income away from wages and toward profits.

Mind you, with regard to equity investments, surely micro specifics will dominate macro generalities. Some sectors and, within sectors, some companies, are going to profit handsomely from the new revolution, while others will not benefit at all or lose seriously from it, even to the point of extinction. The robot and AI revolution will transform the structure of production, and indeed the very things that we produce. But these words could be somewhat misleading. "Produce" isn't quite right, and whatever the appropriate verb is, the object will more likely be a "non-thing" than a thing.

Investments will perform according to how the underlying assets and their management meet the challenges and seize the opportunities that robots and AI bring. To give you a sneak preview of what is to come in Chapters 4–6, I see four boom areas: healthcare, leisure activities, personal development and care of the elderly.

Similar considerations apply to the property world. Residential property will also be buffeted by opposing forces. Any increase in real interest rates will tend to depress prices, while much greater incomes will tend to increase them.

Much the same can be said for commercial property. But here also micro specifics are likely to play an overwhelming role. And commercial property could be subject to a significant negative macro factor. If the amount of time spent at work will be substantially reduced because people choose (or are forced) to take more leisure time (which I will discuss in the next chapter), this will surely accentuate the forces that have recently tended to reduce the demand for commercial office space. Over recent decades, without any influence from AI, or at least not much of it, the long-term trend in the demand for commercial space has been downward. Since 1980, UK real office rents have fallen by 20 percent. Moreover, there are potentially some huge implications for commercial property values from the possible spread of driverless vehicles, which I will discuss at length in Chapter 5.

The macroeconomy of the future

My analysis of the likely shape of the macroeconomy in the Robot Age has thrown up some conclusions that many readers may find surprising:

- There is no convincing reason to believe that the AI economy will be an age of unemployment. It is far from inevitable that aggregate demand will have a tendency to fall short. Indeed, there are good reasons to believe that it could be strong.
- If aggregate demand does fall short, then we should expect macroeconomic policy – including fiscal relaxation and low interest rates – to swing into action to maintain demand and support jobs, at least temporarily.
- Nor is there a convincing reason to expect that the Robot Age will be an age of low inflation. Admittedly, though, as robots and AI come to exert an increasing influence over the economy, this will deliver a series of disinflationary shocks, akin to what happened in the 1990s thanks to

globalization and the rise of China. Moreover, their influence may help the economy to run at higher levels of demand and employment without provoking accelerating inflation.

- The rate of economic growth is likely to increase as productivity growth picks up thanks to the influence of robots and AI. Over time, this will potentially deliver a huge rise in living standards.
- Aside from any temporary policy of low interest rates to counter weak aggregate demand, real interest rates are likely to rise, with rates returning to something like the pre-GFC levels, or perhaps even higher. What happens to nominal interest rates depends upon the inflation regime that prevails in the Robot Age and on that there is no overwhelming case to believe one thing rather than another.
- Higher real interest rates would have a depressing influence on the value of all assets. Except for bonds, for most assets, especially equities, this effect could be offset by the effects of stronger economic growth. That said, micro effects will dominate and many equities will languish even as others go through the roof.

These are a pretty dramatic set of conclusions. But a word of warning is in order. At this stage they must be regarded as provisional. For there is something missing, and that something may have a profound effect on the economic issues discussed in this chapter and the conclusions laid out above. In essence, the spread of robots and the development of AI will bring a significant increase in our productive potential. But what will we do with that potential?

Much of the discussion in the AI literature focuses on the prospect of enforced leisure through technological or demand deficient unemployment. I have argued in this chapter that this is an unlikely prospect. But what about human choice? My contention is that, at the margin anyway, human beings will have to decide how much they work. At one extreme, they will want to carry on working as much as before, albeit in different ways. Accordingly, they will enjoy much higher incomes to spend on goods and services.

At the other extreme, they will want to work so much less that total output and incomes do not rise at all. In that case the robot and AI revolution will show itself not in an increase in GDP but rather in an increase in leisure. In that event, the employment pessimists might even claim that they had been vindicated as "unemployment" rose. In fact, they wouldn't have been vindicated at all because the increase in "unemployment," or underemployment, would be voluntary and it would be a boon to humanity. But the implications for the economy would be quite different from the situation where people carry on working as much as before.

And there are umpteen possible outcomes between these two extremes. So where on this spectrum will our future lie?

PART II

Jobs, leisure, and incomes

4
Work, rest, and play

"Work saves a man from three great evils: boredom, vice and need."

Voltaire[1]

"Choose a job you love and you will never have to work a day in your life."

Anonymous[2]

Most of the techie literature about our AI-driven future, and even a good deal of the writing by some economists (who should know better), envisages a future in which human beings are helpless victims of the onward march of technology. Jobs disappear and people are left on the scrap heap through no fault of their own. And there is next to nothing they can do about it. It is a case of *force majeure.*

By now you should have got the idea that this is not my vision – at least not until the Singularity takes hold – whenever that is. (In this book it appears in the Epilogue.) Indeed, as I hope I have convinced you, in the new robot- and AI-driven world there is no *macro* reason why everyone who wants to work cannot do so.

Of course, particular individuals may be left out and fail to reskill and find work. The transition will be painful. And there will be issues about the income that they can earn from their work. (I will discuss these questions in Chapter 6.) Moreover, the prospect of individuals and communities being left behind, as well as major changes in the distribution of income, give rise to important issues of public policy, which I turn to in Part III. But this is a far cry from the jobless and penurious future that the ultra-pessimists envisage.

The thrust of the vision of the future to emerge from the last chapter is that how much people work in the future will *not* be the result of *force majeure*. Rather, it will be the outcome of choice. Individuals themselves and society as a whole will be able – and indeed will need – to choose the right balance between work, rest, and play. And by "right," I mean the balance that best suits them.

So, what will best suit them? This is a controversial issue, and it is impossible to give a hard-and-fast answer. Some factors and arguments pull strongly in the direction of people predominantly choosing to work just as much as they do now. Others point in the direction of people predominantly choosing increased leisure over increased income. In what follows I will consider both standpoints.

If people do choose, in the future, to spend less time at work, and more at leisure, then two other key questions arise:

- When during their working weeks and working lives will that extra leisure be taken?
- What will they choose to do with their extra time?

This second question is not just a matter of idle curiosity. If they cannot spend this time "profitably," then they are less likely to choose leisure over work. Moreover, their choices will have an effect on the structure of employment.

Once we have answered these key questions then we will be in a position to answer the meta-question posed at the end of the last chapter, namely will the increase in productive capacity that robots and AI will bring be taken out primarily in the form of increased output or increased leisure?

To form a view on these matters we need to delve into mankind's attitudes to work and leisure, drawing on our long historical experience in our own culture in the West and, whenever relevant, on the experience of other cultures. We will be dealing here with issues that are relevant to the human condition in the modern world even without the robot and AI revolution. If the latter proves to be a damp squib, as many cynics presume that it will, we will still face these key questions about work and leisure. But because of the nature of the challenges that the robot and AI revolution potentially pose for the world of work, they are especially relevant to our thinking about the AI economy.

Ancient and modern

Throughout history, mankind's attitude to work has been contradictory. At one extreme, there has been a view that work is the key to a sense of purpose and value, and even the route to godliness. At the other extreme, work has been seen as the bane of life, exhausting, stultifying, enslaving, and even dehumanizing.

Both these attitudes appear in the Christian tradition. "The devil makes work for idle hands to do" is a well-known Christian proverb which may originate with a remark by Saint Jerome in the fourth century. Yet, in the Garden of Eden, work appears to have played no part at all. And this vision of a workless life is not confined to the Old Testament. In Matthew's Gospel, Jesus tells his disciples not to worry about work and money. He says: "Consider the lilies of the field, how they grow: they neither toil nor spin."[3]

Many of the negative attitudes toward work that persist today can be traced back, if not to the Garden of Eden, then at least to the Industrial Revolution. This sucked people into factories, changing the character of work. The division of labor championed by Adam Smith may have been more efficient than one person working at several different tasks, but it also had its downsides, and he acknowledged as much. In *The Wealth of Nations*, published in 1776, he wrote: "The man whose whole life is spent in performing a few simple operations, of which the effects are perhaps always the same, or very nearly the same, has no occasion to exert his understanding."[4]

Marx took this point further. He wrote: "Owing to the extensive use of machinery, and to the division of labor, the work of the proletarians has lost all individual character, and, consequently, all charm for the workman ... He becomes an appendage of the machine, and it is only the most simple, most monotonous, and most easily acquired knack, that is required of him."[5] The result, Marx said, was that workers only felt truly themselves when they were not at work.

So, you could readily argue that mankind's apparently contradictory attitude to work is not a contradiction at all but rather a well-justified difference of view about different sorts of work. From the earliest times, work was mainly physical, hauling rocks and soil, digging, fashioning materials into useful form, bending, carrying, pushing, and pulling. In time, of course, humans learned to employ animals to do much of the backbreaking work of carrying and pulling. But this left much heavy labor still to do.

Where work is actively disliked in the modern world it is usually because it amounts to some form of drudgery, sometimes still involving considerable amounts of physical labor. In George Bernard Shaw's play *Pygmalion*, adapted into the musical and subsequently hugely successful film *My Fair Lady*, Eliza Doolittle dreams of being so well off that she has to do nothing at all, being able just to enjoy putting her feet up. The sentiment is similar to the one expressed in the traditional epitaph for a charwoman: "Don't mourn for me, friends, don't weep for me never, for I'm going to do nothing for ever and ever."

Yet it is clear that the drudgery of work does not necessarily derive from the intensity of physical labor. Indeed, many people have attested to the benefits and sense of satisfaction from physical exertion. As if to endorse this view, now freed from hard physical labor at work, in most of the Western world, men and women now voluntarily chain themselves to machines for several hours a week, and sometimes even several hours a day. Admittedly, the purpose is not the pursuit of money but rather the pursuit of health and the body beautiful. These places of modern enslavement are called gyms. Interestingly, the expression most commonly used to describe this activity is "working out." (Having myself experienced the agony, although not so much the ecstasy, I well understand the use of the word "working." But I have never understood what meaning the word "out" is supposed to convey.)

The widespread dislike of the sort of work available once the Industrial Revolution had taken hold was based, not so much upon its physicality as on its fundamentally dehumanizing character as people became slaves to machines. There is an extensive academic literature on this subject, but perhaps the most vivid presentation of this view comes from the novels of the early twentieth-century

English writer D. H. Lawrence. He saw industrialization as not only besmirching and polluting the countryside but also as something that diminished people and polluted their souls.

Lawrence was strongly influenced by what he saw of the coal industry and the working life of the miners in his native Nottinghamshire. But it is clear that nonmanual work can also feel like drudgery. Many office and administrative jobs can be intensely wearing, and even soul-destroying. What seems to make work display the "drudgery" characteristic is a combination of:

- Repetitiveness.
- Lack of connection with colleagues.
- Lack of connection with the end-product and the end-user.
- Lack of belief that what is being produced/delivered is inherently worthwhile.

If it is thought that work is bound to be like this, then it is not surprising that utopia is envisaged as a world without work. Ironically, in Karl Marx's vision of a communist future, the triumph of the *working* classes was to achieve a world without work – and without the need for it. Capitalism was a necessary stage in economic and social development because it delivered the goods – literally. But as productive potential grew and grew, eventually the need for capitalism, as well as its ability to survive, came to an end. Under communism, human relations and political institutions would be transformed because goods were now available in abundance.

Although they envisaged a rather different route and a rather different end state, other great thinkers reached a similar conclusion. Toward the end of the eighteenth century, Benjamin Franklin, one of America's Founding Fathers, predicted that eventually four hours work per day would be enough. The remaining hours would be for "leisure and pleasure." Later, the playwright George Bernard Shaw outdid this prediction. In 1900 he suggested that by the year 2000 workers would be putting in just two hours a day. Decades later, the influential think tank, the RAND Corporation, forecast that in the future 2 percent of the population would be able to produce everything that society needed.

The work/life balance

And yet, as we shall see in a moment, a good deal of modern work does not fit into the drudgery model at all. Indeed, many people seem actively to like their jobs. As we move into the AI economy, these two strands of opinion about work are very much in evidence. Accordingly, if we are on the brink of an age of abundance without the need for work, this is variously seen as liberating and terrifying.

You will already have noticed the high regard in which I hold John Maynard Keynes on the subject of macroeconomics, which he more or less invented. You may be surprised, though, to find him playing a major role here in the discussion about the balance between work and leisure. Mind you, as will soon become clear, his thinking on this subject is very far from being the last word. Indeed, his contribution in this area posed more questions than it provided answers. But it highlights a key question relevant to the future of work in the Robot Age, namely why do people currently work as much as they do? And whatever the reasons, are they bound to continue in this way in the future?

Interestingly, although he envisaged a rather different route to this end state, Keynes had a similar vision to Marx. In an essay entitled "The Economic Possibilities for Our Grandchildren," published in 1931, Keynes suggested that in a hundred years' time the standard of living would be between four and eight times as high as it was then.[6] This, he claimed, would be enough to end the economic problem, that is to say, shortage, and to replace it with abundance. Hence the question of what to do with our time.

In some respects, this essay of Keynes's is a bit of an embarrassment. It wreaks of the prejudices of the comfortable upper-middle-class intellectual, and its vision of the good life comes straight out of the way of life of Keynes and his friends, particularly the members of the Bloomsbury Group. It combines a total lack of interest in the distribution of income across people and classes with an anxiety about how lesser people, that is to say, people outside his circle, could learn to spend their time profitably.

Yet much of Keynes's essay, written nearly a century ago, reads remarkably fresh today. He wrote:

> *We are suffering just now from a bad attack of economic pessimism. It is common to hear people say that the epoch of enormous economic progress which characterized the nineteenth century is over; that the rapid improvement in the standard of life is now going to slow down – at any rate in Great Britain; that a decline in prosperity is more likely than an improvement in the decade which lies ahead of us.*
>
> *I believe that this is a wildly mistaken interpretation of what is happening to us. We are suffering, not from the rheumatics of old age, but from the growing-pains of over-rapid changes, from the painfulness of readjustment between one economic period and another.*

For this vision of Keynes's to provide us with a way to approach the question of the work/life balance in the AI economy, we first need to interpret what has happened in the period since Keynes wrote his essay. We can then turn our attention to the future.

Was Keynes right or wrong?

The period since he wrote his "Grandchildren" essay has been both kind and unkind to Keynes. Since he first gave this paper in 1928 and, after various revisions, it was published in final form three years later, his hundred years is not up quite yet. But, as I write this, with only just over a decade to go, already the verdict is pretty clear.

First, let's give the man his due. Keynes was thinking these startling thoughts after many years of dreadful economic performance in the UK and as the world economy was about to descend into the Great Depression. At the time, the idea of living standards rising to between four and eight times the then present level within a hundred years must have seemed a fantastical vision.

Yet this apparently wild vision of material progress has come to pass – at least in the developed world. Depending upon exactly which year you take as the starting point and which definition of "standard of living" you take, in the USA and the UK, the standard of living in 2018 was between five and a half and seven times what it was when Keynes came up with his vision. And, at the time of writing, there is still another 10 years to go to complete Keynes's span of a hundred years. (Interestingly, in his "Grandchildren" essay, he had nothing to say about the *un*developed world, or what we would call today "the emerging markets." In what follows here, I, too, will focus on the developed economies. But I will bring the emerging markets in shortly.)[7]

Yet, although Keynes was broadly right about the overall increase in average incomes and living standards, the end of work, or even the reduction in the working week to the 15 hours that he envisaged, has not happened. Most adults are currently engaged in full-time employment for most of their lives, working for 30–40 hours a week and, in many cases, much more. So the problem of what to do with all that leisure time has simply not arisen.

Moreover, some key features of the modern labor market are the very opposite of what Keynes envisaged. Indeed, many people in the professions seem to be working longer hours. Stockbrokers used to lead lives of considerable ease, with their normal working hours running to about six or seven per day, at most. And a good deal of that time was devoted to "lunch." Nowadays, by contrast, financial professionals are often at their desks by 7am and work every hour that God sends – and without "lunch."

Similarly, although they typically start later in the morning, legal professionals also work very long hours, often late into the night. (Quite what explains the different diurnal/nocturnal balance of work between financial professionals and lawyers is beyond me.) This is a far cry from how the legal profession functioned during the nineteenth century, and indeed during the twentieth century up to the 1980s. Until comparatively recently, a good few successful English lawyers were able to conduct their legal business in the morning and then repair to the House of Commons for the afternoons and evenings.

Speaking of politics, in the nineteenth century British Prime Ministers might take themselves off to the Continent and/or to their country estates for months over the summer. Now they are supposed to be hard at it pretty much all year round. And woe betide them if they happen to be caught sunning themselves abroad when some national disaster strikes. (Interestingly, this greater intensity of effort – or at least the appearance of it – does not seem to have produced any obvious improvement in the quality of government.)

Ordinary office workers also typically toil for longer than they used to, often with considerable commuting time on top of their formal office hours. Moreover, during their long commutes they might well spend quite a bit of time working on their laptops or smartphones, and they might well continue doing this at home, during their supposedly leisure time.

Meanwhile, the number of women working outside the home has risen enormously.[8] Nowadays, in most countries of the developed world, both members of a couple work outside the home, in contrast to the old model under which one person, usually the woman, stayed at home and looked after the children and the household. The result of this change is that for many people, when they get home after work, this is far from being an elongated period of leisure. Rather, it is just another sort of work. For all these wage and salary slaves, Keynes's vision of a life of leisure and the associated challenge of what to do with all that time seem to come from another planet.

There is even some evidence to suggest that many people – although not in the professions – would like to work *longer* hours than they actually do. In the USA and the UK, where workers have some ability to choose their hours, hours of work far exceed what is normal in European countries where union policies and legal restrictions limit working hours. The USA has a per capita GDP about 30–40 percent higher than France or Germany but over the course of a year the average employed American works about 30–40 percent more hours than the average employed person in those countries. What's more, the gap between America and Europe has been increasing. In the USA average weekly hours worked by people of working age was 24 in 1970. In 2004 the figure was 25.[9]

What explains this apparent modern preference for work over leisure, at least in the Anglo-Saxon world? Only once we have some sort of answer to this question will we be able to form a view on the future work/life balance in the AI economy.

Economists to the rescue?

One straightforward economic explanation as to why people have carried on working so much despite increasing incomes is that "the substitution effect" (arising from the higher opportunity cost of leisure, thanks to higher incomes from work) has outweighed "the income effect," which could be presumed to increase the demand for leisure.

This is not just an economic answer; it is an economist's answer. (And in my book, that is not necessarily a mark of approbation!) Putting this supposed explanation into plain English, the fact that people now earn more from work favors the choice of work over leisure, while the fact that people are better off allows them to take at least some of the benefit in the form of more leisure and less work. From the evidence, we can observe that the former effect has outweighed the latter.

This describes the phenomenon, but I am far from convinced that it properly explains it. Arguably, technological progress has greatly increased the subjective value of leisure by expanding the range of entertainment possibilities. By contrast, in the preindustrial era, "leisure" must have involved a fair amount of sitting in the dark.[10] This factor greatly favors the choice of less work and more leisure now. So, we are left still looking for an explanation for our apparent choice of work (and income) over leisure.

The competitive urge

In fact, we don't need to look very far to find some powerful forces keeping people's noses to the grindstone. Perhaps the most important is the competitive instinct. Even if we reach a point where, in an absolute sense, earning extra income is not much prized because the extra "things" that can be bought with it are not much needed or desired, competition with one's neighbors and peers can still propel intense effort. People may still want to show that they are at least equal, or perhaps superior, to their comparators, whoever they may be.

They may even feel, if not quite a sense of competition with their earlier selves or their parents, at least a sense of wanting to show demonstrable progress against their own, and their parents', former lives.

It is possible that increased *inequality* (if, indeed, there has been an increase in inequality – see the discussion in Chapter 6) may also have contributed to the choices made between work and leisure. Those at the bottom of the pile want to work in order to acquire what those higher up the pecking order have acquired. Meanwhile, those at the top are keen to maintain their margin over those below.[11]

Interestingly, Keynes foresaw the importance of relativities. In his "Grandchildren" essay he wrote:

> *Now it is true that the needs of human beings may seem to be insatiable. But they fall into two classes – those needs which are absolute in the sense that we feel them whatever the situation of our fellow human beings may be, and those which are relative in the sense that we feel them only if their satisfaction lifts us above, makes us feel superior to, our fellows. Needs of the second class, those which satisfy the desire for superiority, may indeed be insatiable; for the higher the general level, the higher still are they.*[12]

There are several drivers of the desire for relative success. The first is purely economic. However rich our society becomes, some things will always be in strictly limited supply – the best seats at Wimbledon or Covent Garden, the apartment with the best view of Central Park, or the house in Saint-Tropez with the best view of the ocean. These are what economists call "positional goods." Since, by definition, society cannot produce more of them, their pursuit by individuals will result in a constant struggle for economic achievement. A particular individual may succeed in bagging these goodies but only by denying them to others.

The other key drivers are in essence psychological and societal. The first is that most human beings just like winning. The drive to gain promotion, achieve a higher position and earn more money can derive purely from the desire to "win," to "beat the others." This is particularly true of some successful businesspeople who seem to have an insatiable desire for work. Admittedly, this is sometimes because they really enjoy what they do, but it is often because wealth and obvious material success are a means of "keeping score." And they desperately want the score to come out in their favor.

Another (related) driver is the pursuit of power. To a considerable degree, power is associated with relative – and note, not absolute – economic success. In modern societies, this power is most in evidence, and most pursued, within companies and corporations, but it is also to be found throughout society, including in schools and universities. This power motive is closely connected with another driver behind the desire for relative success, namely the pursuit of status.

The forces of competition for relative success, driven by all the factors mentioned above, are very clearly at work in professional services businesses – in New York, London, Hong Kong, Singapore, Mumbai, and indeed anywhere operating in a global professional environment. Bankers, lawyers, accountants, management consultants, and many other professionals toil for long hours in order to make their way up the corporate (or sometimes the partnership) ladder.

Law firms have traditionally operated the so-called "lockstep" model, under which pay and status are directly linked to longevity in the firm. A junior lawyer joining one of the top firms will usually work like stink in order to "make partner" after perhaps eight or nine years. While they are trying to achieve their objective, they will typically work about 100 hours per week – roughly three times the number of hours laid down as the maximum under France's 35-hour week law. Generally, fewer than 10 percent of these strivers will reach partnership level.

Enjoyment and reward

The drive for relative success and the pursuit of "victory," power, and status explain the phenomenon of continued heavy work schedules in a rather negative way. But there is another explanation that reflects a more attractive side of human nature. After the basic, compelling, need to work for survival, and then subsequently the drive to secure comfort, subsided in importance once these objectives were largely achieved, human beings have increasingly looked to work as a source of comradeship, diversion, interest, entertainment, and purpose.

These are very real and powerful drivers. And the flipside is also extremely powerful. Umpteen surveys have reported that unemployment is one of the biggest causes of self-reported unhappiness, over and above what can be put down to mere loss of income.[13] In other words, work can be *enjoyable* – and its absence can be downright awful. In many cases this isn't so much because of the innate nature of the work as the social interaction with colleagues that it brings. For instance, in America between 40 and 60 percent of workers have dated a fellow employee.[14]

Furthermore, over and above the fun aspect of working, there can be deeper positive feelings, too. Work can generate a sense of pride and identity and give people a sense of purpose. This is true for both employees and employers. And for entrepreneurs, arguably there are even higher sources of pleasure – the sheer excitement of creating an enterprise.

The idea of work as the source of meaning and value reaches its apogee in the concept of "the career," that is to say the notion of an upward progression that leads to more money, status, importance, and worth over time as the person ascends the ladder, corporate or otherwise. I suppose there were always people whose working lives had this aspect to it – in the Church, or the army, perhaps. But the notion that many, let alone a majority of, working people would have "a career" is a recent phenomenon. Most people had a living, a job, an occupation, an office, or they owned a business or, if they were lucky, had a calling, but not a "career," with its sense of going ever onward and upward.

Nowhere does any of this feature in Keynes's "Grandchildren" essay. But Keynes is far from being alone among economists. Trapped in their world of Stygian gloom, economists have often been guilty of having largely missed the enjoyment factor. They have typically seen work as an unmitigated bad that people only engage in because of the money that they are paid, as reflected in the frequent use of the term "compensation" to refer to salaries and employer-provided benefits. Yet, although toil and drudgery may well have been many people's experience of work, at least until recently, and still is in some parts of the world, economists' notion of work as an unmitigated horror is well wide of the mark. Among other factors, they have been unduly influenced by the Marxist concept of alienation.

Rediscovering the case for leisure

Given these powerful points, when thinking about the future of work it is easy to swing from one extreme to the other – from the Marxist/Keynesian belief that given mass affluence, working hours are set to plummet, to the belief that the work ethic is so strong that full-time work is endemic to human nature and to the human condition. If you take the latter view, then you will believe that the vision of greater amounts of leisure in the future was always going to prove to be a mirage. And this extends into the Robot Age, leading to the presumption that, even when robots and AI do most of the "heavy" labor and incomes are much higher, people are going to want to work just as much as they ever did.

Yet I think this *workist* view is overstated. It is high time that a more balanced approach got a proper hearing. So, I am now going to look at the other point of view. The case against the complete rejection of the Keynesian vision that more leisure is the way of the future rests on several strands of evidence that I will review in turn:

- The facts about past trends in working hours.
- An alternative economic explanation for why Keynes's prediction of reduced working hours has not been fulfilled.
- Evidence that, contrary to the discussion above, for many people work is *not* enjoyable.
- The idea that current working hours are not in accordance with individuals' underlying preferences.
- Possible changes to the sources of status.
- The possibilities for achieving a sense of meaning and purpose outside work.

Historical evidence

Have we always worked so hard? Of course, we have no data on how hard people worked in primitive societies. In any case, working hours must surely have been dictated by the climate and the seasons. Once societies became settled, over the winter months people probably did little or no work and then, when spring came, they would have worked during most, if not all, daylight hours.

Actually, we do have the evidence of less developed agricultural societies today to consider. In those societies people on average do not seem to work the same number of hours that are common in Western society today.

Interestingly, there is some evidence that, in preindustrial Europe, people enjoyed very extensive holidays, albeit not usually involving foreign travel. The Harvard historian Juliet Schor has estimated that in Europe around 1300, because the calendar was packed with holy days and feasts, holidays accounted for at least one-third of the year. In France, the proportion of holidays was about a half. (This last point goes to show that some important characteristics continue across the ages.) Schor has written: "Our ancestors may not have been rich, but they had an abundance of leisure."[15]

Turning to more modern times, Keynes may have been wrong about recent years but over the long haul he was dead right. There has been a dramatic reduction in average working hours. Between 1870 and 1998, in the highly industrialized countries the number of hours worked per annum per employee fell from 2,950 to 1,500 – a virtual halving. OECD figures for the years after 1998 reveal a further decline (although, as with the earlier figures, less so in the USA).[16] In the UK, the average working week has declined from about 59 hours in the mid-nineteenth century to about 32 hours now.

Not as well-off after all

There is even a straightforward economic explanation for why in recent decades people have continued to favor work over leisure to a greater extent than some people, including Keynes, might have anticipated. Until recently, material improvement only made available to all the very basics of life – decent and sustaining food and drink, heating and shelter, clothing, mobility and entertainment. As people became better off, this hardly meant that they were equipped, or paid, for a life of leisure and ease. It simply meant that they could now aspire to the next basic material need.

After all, as late as 1967, almost 14 percent of homes in England and Wales were not equipped with an indoor flushing toilet. In 1960, 95 percent of homes in England and Wales did not have central heating. And the figure was still over 50 percent in 1976.[17] It is only once these improvements in the basics have been secured that human beings can reasonably contemplate preferring increased leisure to increased income. But the implication is that once these basic material needs have been fully met, as they are beginning to be now, at least in the advanced economies of the West, then people may well start to choose more leisure over work and income.

You can see something similar to this in current trends in retail spending. Across a broad range of goods and services people have in recent years tended to devote increased spending to improvements in quality rather than to increasing numbers or quantity. This applies to food, furniture, and cars. It is perfectly plausible to assume that once they have achieved a decent amount of quality improvement in regard to the basics of their lives, people's preferences will turn toward more leisure.

Admittedly, there is a qualification. Economic growth does not merely make available more of the things that people already have, or even better quality versions of the things that they already have, but it also produces completely new goods. People now aspire to have these new gizmos – televisions, cars, washing machines, computers, smartphones, tablets, and so on and so forth. Mind you, this only gets you so far because what were once "must-haves" quickly become ubiquitous. Moreover, their price usually falls dramatically in the first few years after their introduction. Accordingly, the need to acquire the latest gizmo does not provide that much of an incentive to keep working all hours – until the next gizmo comes along.

Actually, for the USA there is a very simple economic explanation of why Keynes's prophecy has not been fulfilled. Over recent decades, most workers' wages have not increased. For American males in their thirties, median real wages were lower in 2004 than they had been in 1974.[18] (The reasons for this are discussed in Chapter 6.)

The anti-work preference

We should not blithely accept at face value the "work is fun" explanation for continued long working hours that I discussed above. There is a serious risk of mistaking the fulfillment of some people at work for the experience of everybody. For many people, work isn't always quite what it is cracked up to be. In fact, many, if not most, people are alienated by their jobs – still. They find them tedious, meaningless, and boring. According to Deloitte's *Shift Index*, 80 percent of people hate their job.[19]

And there is a fair bit of evidence to suggest that in countries where people work long hours, they are less happy than their equivalents in countries where people work fewer hours. In South Korea, for instance, people work 2,232 hours annually, some 473 more than the OECD average. They report low levels of happiness.

At the other extreme lies Denmark. Danes work on average 1,595 hours a year, some 200 hours fewer than the OECD average. Denmark regularly comes out as one of the happiest places on earth. These are not isolated examples. Other countries with low numbers of hours worked – Sweden, Finland, Norway, and the Netherlands – also come out as "happy." Equally, in countries where the number of hours worked is high – Greece, Poland, Hungary, Russia, and Turkey – people are more miserable.[20]

Moreover, the developing subject of "Happiness Economics" suggests that, beyond a certain point, extra income provides little extra happiness.[21] And, as if we did not already know this, it is also revealing that happiness is associated with the strength of human relationships, particularly with close family and friends. And yet the pace and intensity of modern life constricts the time available to spend with close family and friends and threatens to undermine these relationships.

Are today's working hours a reflection of choice?

There is a fundamental problem with such information that we have on people's preferences between work and leisure. In the literature it seems to be widely assumed that the number of hours that people work reflects their free choice, given the rewards, and the various alternatives. Yet, it is far from obvious that this assumption is justified.

Actually, the facts are a bit more complicated than the averages on working hours suggest. Interestingly, the age-old inverse relation between hourly pay and hours worked has reversed. In the UK, men in well-paid full-time employment are, on average, working more hours than they did 20 years ago. Meanwhile, men at the bottom of the earnings scale, but still in full-time employment, are working fewer hours. Additionally, many more low-paid men are working part-time.

Apparently, this does not accord with people's preferences. UK official data show that 3.4 million people want to work more hours while 3.2 million want to work fewer hours for less pay. Those wanting to work more are typically low-paid waiters or cleaners. Those wanting to work less are typically highly paid doctors or other professionals.

If you are a successful professional, it is difficult to choose more leisure over work. It is often a case of all or nothing. For instance, it is rarely open to an individual lawyer to alter the terms of his or her employment to reduce the number of hours although, increasingly, female lawyers are leading the way in demanding a better work/life balance. But swimming against the tide is extremely difficult and opting to work fewer hours in these thrusting professional services firms usually comes with a heavy price in retarded career development.

And this is a hard thing to do. Humans are social beings and they respond to social pressures. As we noted above, over the last 50 years, to a much greater extent than was true previously, success at work, and even long hours spent at work, have become overwhelmingly the most important indicators of status.

Moreover, there is a coordination problem. Extra leisure time may only be especially valuable to you if you can share it with family and close friends – and they probably feel the same way. But we all bargain separately and make our work/leisure choices, if we can at all, separately. Accordingly, it would be possible for everyone to prefer more leisure, provided that others also had more leisure yet, because of the coordination problem, for us instead to accept a combination of more work and less leisure than we would like because "we don't want to be out on our own."

The key to changing these work practices lies with wider society. Attitudes and social norms are critical. But there may be ways in which government and corporate actions could make a difference. For instance, the structure of employment benefits makes employers keener to employ one person to work overtime than to employ two people working the same number of hours.

The changing sources of status

In the above discussion about why people in the West have continued to work long hours, despite being much better off, I made much of the influence of relativities and considerations of power and status. Yet the drive to work coming from the pursuit of relative position, power, and status isn't bound to be so compelling in the future. If the pursuit of relative success is a game that you do not care to play, then you can opt out of it. And in our society some people do, consciously choosing to abandon "the rat race" and thereby accepting a lower material standard of living in pursuit of greater contentment.

Similarly, power is currently connected with money because of the way our society and political systems are structured. But things are not bound to be this way. In previous societies, power derived, in part at least, from birth. And even in today's societies political power does not always derive from, nor is it necessarily associated with, great wealth.

It is similar with the pursuit of status. Here too the ancient world was different. In many parts of the ancient world, heavy work was performed by slaves. In ancient Greece, for instance, production appears to have been dominated by them. In Sparta there was a slave tribe. These were the Helots, who were a conquered people who were kept alive to work and breed more slaves. The native Spartan men did not engage in work at all. Their role, and their source of dignity, was to engage in martial activities, both training for war and waging it.

In ancient Rome too, slaves performed all the heavy-duty work – actually, not only the heavy-duty work but just about all work, both in domestic service and, for the more senior and more favored ones, tending to all the affairs of their masters. It has been suggested that in both ancient Athens and ancient Rome up to a third of the population were slaves.[22]

Interestingly, in neither ancient Greece nor ancient Rome did status derive from work as such. Instead it derived mainly from birth, wealth, martial prowess, and public office. Often these were closely related. Wealth predominantly derived from the ownership of land, which was inherited, while the ability to attain senior military rank or political office was also closely associated with wealth and birth. Much the same seems to have been true of ancient Egypt, and most ancient societies in Asia.

The societies of medieval Europe were not much different, except perhaps that we can add the church to the list of sources of status. At some point, even before the Industrial Revolution, there was more scope for wealth to be earned through commerce rather than simply through the ownership of land, and, in some cases, for status to be achieved as a by-product of wealth, although in much of society this was widely frowned upon. Indeed, in some quarters it still is. "Old money" carries a cachet that the "nouveaux riches" may long for but which, by definition, they cannot attain – at least not for several decades, if not centuries.

These points about the sources of status have a bearing on one of the key drivers of the pursuit of relative success, discussed above. The competition for positional goods and power are not bound to be all-conquering. After all, they both occur partly because they confer status. But in the new world status *could* derive primarily from other sources: being beautiful, being a well-rounded and developed person, being successful at sport, or whatever.

Purpose and meaning

Even if many people don't need work as such, or the money that comes from it, or even the power and status, they might nevertheless need a sense of *purpose* that work currently provides. Yet purpose does not need to be supplied by what we call "work." Specifically, people do not need to use most of their waking hours in pursuit of money in order to have a purpose. Purpose and meaning can be given by a wide range of things: the pursuit of a hobby or sport, or the desire to create something, or to improve one's proficiency at some skill or practice. This can apply to bridge, golf, skiing, speaking foreign languages, and a host of other things. Fulfillment and purpose could also come from charity and community work or meeting extreme challenges, such as running a marathon or climbing Mount Kilimanjaro.

Even this apparent need for purposive activity could be cultural rather than inherent. There are some non-Western cultures, including some in the South Pacific and parts of Africa, where people don't seem to be bothered to do much work across the year. Their status structures do not seem to have suffered, and people do not seem to be afflicted by a lack of meaning and purpose in their lives. And, of course, the upper classes in all European societies traditionally did not work at all but rather amused themselves with sports, hobbies, and various social activities. What gave them a sense of purpose? Perhaps they didn't need one?

Interestingly, in the Gulf States today the working habits of the rich Emirati are quite similar to what Keynes had envisaged in his "Grandchildren" essay. They typically work three or four hours a day. Actually, their life circumstances are not dissimilar to those of the rich English aristocrats who probably influenced Keynes's thinking. Most notably, the rich Emirati have considerable private wealth that provides them with an income even if they do very little work.

How could shorter hours be achieved?

So, the upshot is that I suspect that for many people in the advanced economies there is a strong latent demand for more leisure time. I feel sure that, as people get richer, except for those on low incomes who, in any case, often work few hours currently, there will be a marked desire to reduce working hours.

Suppose that I am right and many, if not most, individuals in advanced Western societies will choose to work fewer hours and to enjoy more leisure time as they get richer over the coming decades. And suppose that the institutions of society, including employers, are inclined to facilitate this. Where in the working lifetime would the extra leisure time come? There are six major possibilities: a shorter working day, a shorter working week, longer holidays, a shift to more single-earner households, longer periods in education, and longer periods in retirement. I briefly consider each of these below.

A shorter working day

Normal working hours are today widely regarded as something like 9 until 5.30 but this in not cast in stone. Actually, for the customer who wishes to visit a shop, speak to someone in an office, to have some service provided, or to get to some task accomplished, these "normal working hours" can be very inconvenient, if not onerous.

Clearly, it would be against the thrust of the developments being discussed here for normal working hours to be extended beyond the conventional span. But there is nothing against longer "opening" hours combined with each worker spending fewer hours in the shop, office, school, or whatever. For example, normal "opening hours" could be extended to something like 7 until 7 but with staffing being provided by two six-hour shifts. This would make the normal working week 30 hours if five days were worked, or only 24 hours if four days were worked (see below).

A shorter working week

Another, noncompeting, possibility is to reduce the numbers of days worked. There is nothing sacrosanct about a five-day week, with two days off as "the weekend." Although the Sabbath has been well established as a nonworking day for ages, making a six-day working week the norm, the two-day weekend is a relatively recent concept.

It was once normal for banks to be open on Saturday mornings. In Jerome K. Jerome's charming and much-loved novel *Three Men in a Boat*, published in 1889, one of the eponymous three, George, can only join the others for their weekend further upriver because he "goes to sleep at a bank from ten to four each day, except Saturdays, when they wake him up and put him outside at two."[23]

Moreover, in the Muslim world today, some people work a six-day week, taking only Friday, the Sabbath day, off. In the Gulf States, for many expats this is made up for by the fact that, having started work early in the morning, they can stop work at lunchtime, and often take themselves, along with their families, to the beach.

In Europe, workers particularly appreciate being able to attach a public holiday to a weekend so that they can enjoy a long weekend of three days, or even four. If four-day working were the norm, then such extended weekends, whether spent at home or away, would become normal also. What's not to like?

Actually, things are already moving rapidly in the direction of a shorter working week. The German union IG Metall recently reached an agreement with employers in the German metals and electrical industries on a 28-hour week for 900,000 of its members.

In 2018 the New Zealand insurance company Perpetual Guardian adopted a four-day working week and in the UK the Wellcome Trust has considered moving all its staff onto a four-day week, by giving Fridays off to all 800 of its London-based employees, and without reducing pay.[24] (However, these plans have recently been dropped.)

Extending holidays

Until recent times it was not normal for workers to have paid annual holidays. And many people running their own businesses today still don't. They work all the time. Originally, holidays were the holy days, that is to say, the various religious festivals. More recently, these were added to by various other nonreligious public holidays, such as Labor Day in the USA, or August Bank Holiday in the UK.

When paid annual holidays were introduced, the norm was for there to be only two weeks. For many senior people in Europe this has now reached six weeks, and sometimes even more. For ordinary employees the norm is four weeks.

This state of affairs is not inevitable. The reality of short holidays usually comes as a massive shock to young people leaving university, who have been used to having about half the year with no formal work obligations. Some of them never recover from the shock.

There is ample scope for people to take longer holidays from work. This is particularly true in the USA where two weeks' vacation, or at most three, is the norm. The average American works 30–40 percent more over the course of a year than the average European. On average Americans don't take four days from the two weeks of vacation that they typically receive, contrasted with Europeans who usually take all of their four- to five-week vacations.

If Americans simply switched to normal European working hours and took the number of holidays that the average European does, this would cut US labor supply by the equivalent of a reduction in the US labor force by some 20 million people, out of a total of 160 million.

Dual-earner households

Increased material wealth has been accompanied by an increase in the prevalence of both parents working. There is a good deal of evidence that this has led to much angst and stress, for both parents and children. To some degree, of course, this shift has occurred because of perceived material necessity. But to a large extent it is driven by considerations of status and position in society. As people get richer, there is surely ample scope for this shift to go into reverse.

Why should it be normal for both partners to work full time when there are young children at home – or even if there are not? One way in which people, on average, could reduce their working hours would be if it again became more common for one member of a couple to spend most, or even all, of their time at home, where they might engage in some part-time work – or not. (If something like this does transpire, how the out-of-home workload and responsibilities for home and childcare should be shared between the two parents is beyond the scope of this book.)

Similarly, in most Western societies, it has become normal for mothers, and in some cases fathers, to take several months off following the birth of a child (again, not in the USA). But there is clearly scope for this period to be lengthened.

Longer education

Two other ways of reducing labor supply are available. First, young people could delay getting into full-time employment and could spend longer in full-time education. Already in Germany it is not uncommon for students to delay their entry into the labor force until their mid- to late twenties.

Or, instead of abandoning education at the age of 18 or 21, people could spend periods of time in education during their normal working lives. If you like, adapting what happens at present in the educational world, people could take periodic "sabbaticals" from work.

Whether or not these changes to the educational "timetable" become widespread and, indeed, whether it is a good idea that they should do so, is a subject that I take up in Chapter 8.

Longer retirement

At the other end of our lives, it would be possible for people to retire earlier and thereby to enjoy a longer period of retirement. Of course, at the moment the pressures are all the other way, in response to increased longevity and various pension problems. But that is because our thinking is still trapped in the economics of shortage, including financial shortage. If we accept that in coming decades, robots and AI are going to be able to relieve us of much of the burden of work while also making us richer, then this will be a very different world from the one that we inhabit now and which is driving so much public policy.

Admittedly, increased longevity is now increasing most people's likely period of retirement well beyond what could have been envisaged even a few years ago. Some analysts and commentators are now seeing the advances of medical science stretching normal human life well beyond 100.[25] If the increase in average lifespans continues, never mind if we find a cure for cancer and diabetes, this will potentially cause major issues for society and for our concepts of a normal working life and retirement.

To have both a large extension to life expectancy and a shortening of the normal working life at the same time might be going too far. But at least we should be wary of assuming that as our lifespan extends it is inevitable that our working lifespan should expand commensurately. There will be room for a bit of both: a longer working life and a longer retirement.

A life of leisure

Now we come to the second big "work/life" balance question that I raised earlier. If work were to take up less time, what would people do with all the hours freed up? As so often, there isn't a black-and-white answer. Today many people do not find it a challenge at all to occupy themselves fully. Indeed, many retired people feel very happy and say that the most difficult period of their life was in the middle, when they had young children, mortgages, and career development to grapple with.

Some retirees even say that they don't know how they ever found time to go to work. A combination of bridge and golf (if you can stand it) can be made to take up most of the week. And if that's not enough there is always voluntary work. According to the Dutch historian Rutger Bregman, the countries with the shortest working weeks also have the largest numbers of volunteers and the most "social capital."[26]

Yet there are many people who *do* find not working difficult. Oscar Wilde once said: "Work is the refuge of the people who have nothing better to do."[27] As long ago as 1964 the great science fiction writer and visionary Isaac Asimov envisaged that the death of work would cause serious emotional and psychological consequences. What most worried him was the prevalence of boredom. He suggested that by 2014 psychiatry would be the largest medical specialty. And plenty of people in Western societies do experience an acute sense of diminished status and self-esteem, as well as loss of purpose, when they retire. In many cases, the unhappiness that this produces does indeed lead them to seek psychiatric help.

But it would be wrong to jump from this fact to the conclusion that we face a mass outbreak of unhappiness as virtually the whole population experiences, throughout their lives, what a considerable number of people experience today when they retire. For the experience of today's retirees derives from the fact that they have spent most of their lives working and that, when they retire, other members of society, both those they are acquainted with and more distant members, are currently earning income and enjoying the benefits of comradeship, interest, sense of purpose, power and status from working.

So their experience does not provide any serious guide to how people in general might feel if they did next to no, or at least little, paid work during their lives, or if they experienced long periods of not working, or of working much reduced hours if, simultaneously, most other members of society were in exactly the same position.

How to cope with greater free time and how to make the most of it will surely be one of the tasks of education in the Robot Age. This is something that I will address in Chapter 8.

A view of the future

This analysis has highlighted a major divergence of view about humanity's preferences between leisure and work as people get richer. On one view, much reduced hours of work and increased leisure will be a boon and a liberation. On another, it would not be what most human beings would naturally choose and it might even cause them great angst, as well as producing mass recourse to professional psychiatric help, accompanied by various social pathologies.

To the extent that the first view is correct then, if the AI revolution does indeed produce a massive drop in the demand for labor, or if people are able to choose shorter working hours and more leisure time, provided that people have an adequate source of income – which may need some intervention by the state (to be discussed in Chapter 9) – this should cause no problem for humanity whatsoever. In fact, quite the reverse.

But to the extent that the second view is correct then a collapse in the demand for labor could potentially create huge problems for both individuals and society. Of course, if people's preference for work over leisure is so strong, then they may want to continue working even at very low wages. In that case, the end result may be that a considerable number of people remain in some sort of job for most of the time, albeit on low pay. Then, again, some state intervention to affect the distribution of income may be on the cards.

Actually, we will never know the answer to this question unless and until we are faced with the circumstances that give rise to the question. And if the analysis that I put forward in Chapters 2, 3, and 4 is correct, we never will be, because the advance of robots and AI will *not* produce a dramatic collapse in the demand for labor. (At least, not until the Singularity befalls us. Just to remind you, in this book this happens in the Epilogue.)

It is easy to believe in either of the two extreme outcomes: people carrying on working as much as before, with no increase in their leisure time, and people taking all of their extra productive capacity out in the form of increased leisure time, with the result that there is no increase in GDP per capita or in material living standards. In practice, though, most people will surely want to choose an intermediate position – if they can. Where they in practice choose to be on the spectrum between these two extremes will depend partly upon what happens to their incomes, a subject on which I have largely suspended comment until I discuss it in the next two chapters.

If the result of the changes wrought by the robot and AI revolution is that the income of the broad mass of people falls, as incomes are redistributed toward the owners of capital and the highly skilled, then the broad mass of people are most unlikely to want to supply less labor. Indeed, they might even choose to supply more in order to maintain their standard of living.

Yet this is far from being the inevitable result. Indeed, in the next two chapters I shall argue that it probably won't happen. So, let us suppose for a moment that the income distribution remains broadly constant, with or without intervention from public policy, which I discuss in Chapter 9. What then?

In many parts of the world, including most of Africa as well as China and India, on average people have nowhere near reached the point of material satiety after which continued effort to secure more material wellbeing denotes some pathological condition. Indeed, about 50 percent of the world's population lives on less than $2 a day.[28] As productivity growth offers the chance for increased material living standards, people there will surely overwhelmingly choose to take this opportunity rather than to take out the improvement in their circumstances in the form of increased leisure.

And, to a lesser extent, the same applies to people at the bottom of the income pile in advanced Western societies. These people too are well below any level of income that can reasonably be described as providing full satiation of material needs. Indeed, since many of them currently work decidedly short hours, quite a few might want to work more rather than less – if only the economy can generate enough employment opportunities for them.

But for people in the middle and upper reaches of the income distribution in advanced countries things are quite different. For them I envisage an intermediate outcome. The competitive urge, the pursuit of power and status, and the competition for positional goods will be powerful factors for maintaining something like the current "work/life balance." And Keynes was wrong to see work so negatively, and he underestimated the forces, many of them to do with relative position, and the search for power and status, that drive people to go on working. Equally, he massively underestimated, or even ignored, the importance of distributional factors. The fact that increased leisure needs money to be able to enjoy it to the full will pull in the same direction.

But for a large number of people in Europe and North America, as well as in the prosperous parts of Asia, very long hours of work have become the expression of a social pathology. For the richer members of society to work such long hours is historically anomalous. And it is not healthy. There is ample scope for this to change. I believe that it will. It will become normal for people to work a shorter working day and a shorter working week, with the three-day weekend becoming the norm. Moreover, the average length of holidays will increase – even in the home of nonstop work, the USA.

I am not so sure about longer periods spent in education, as I discuss in Chapter 8. And longer periods of time in retirement, if it happens at all, will come because of greater longevity rather than because large numbers of people choose to shorten their working life. Hence it will not serve to reduce the supply of labor. Indeed, the tendency may well be toward a longer working life, thereby tending to increase labor supply.

Leisure is something that people want more of as they get richer, particularly when, as now, technological improvement has greatly increased what they can get out of it. So, I expect that the world, or at least the developed part of it, will gradually choose more leisure, although nothing like as much as Keynes suggested all those years ago.

The overall effect of these changes will be to reduce the effective supply of labor. This will be made possible by, and will also itself lead to, a shift in the indicators of status. Being a complete workaholic banker, lawyer, or accountant may come to be seen more as a mark of failure than of success.

If I am right about these changes, there will be not only a reduction in the supply of labor but also radical changes in the structure of the demand for labor. As I argued above, the more that some people choose to take more leisure, the more employment opportunities will arise to meet their needs. These employment opportunities will be for humans rather than robots. So the balance of demand for robots and AI versus humans will shift in favor of humans, thereby attenuating whatever erosion there might otherwise be of the overall demand for human labor. In the AI economy, an increasing share of the jobs available for humans will be in the leisure sector – among all the others that are set to surprise us.

5
The jobs of the future

"We are in the midst of a robot invasion. The machines are now everywhere and doing virtually everything."

David Gunkel[1]

"The jobs of the future don't exist today and the jobs of today will not exist in the future."

Stuart Armstrong[2]

Let us begin with brutal honesty. We simply don't know what sort of jobs will be available in the future. After all, imagine yourself in the year 1900 peering into the future. How could you know then that the proportion of people employed in agriculture in the USA would fall to a twentieth of what it was then? Or that all those umpteen thousands of people employed doing things with horses – buying and selling, keeping, stabling, feeding, and cleaning them – would be redundant? Or that there would now be more people employed as mental health nurses in the UK's NHS than there are sailors serving in the Royal Navy? Or that large numbers of people would pay good money to personal trainers to put them through their paces and ensure that they suffered the requisite amount of agony?

History is full of people who have made long-term predictions and who have been proved utterly wrong. Apart from the Reverend Malthus, whom we met in Chapter 1, among economists one of my favorites is the great William Stanley Jevons, one of the most distinguished economists of the late nineteenth century. In 1865 he predicted that industrial expansion would soon come to a halt due to a shortage of coal. Poor old Jevons.

There is a common thread running through the forecasting failures of most experts – or rather two. First, they underestimate the ability of human beings, and the economic system, to adapt. Second, aside from the one big change whose effects they are trying to foresee, they envisage the future as a straight continuation of the past. In other words, they are markedly lacking in imagination.

These failings are not restricted to would-be visionaries past. They can also afflict us. So, we must tread extremely warily. Having said that, and having dosed ourselves with lashings of humility, and drunk deep from the well of skepticism, there is a lot that we can say about the future of employment in the new robot- and AI-dominated future.

Some studies have tried to estimate the number of jobs that will disappear in particular sectors and to hazard a guess at what number of new jobs that will appear. These exercises can have some value. Indeed, I will refer to some of their results below. But such studies are replete with arid banks of dubious numbers. By contrast, what this chapter tries to do is to concentrate on the principles underlying job destruction and job creation in the new economy. It tries to identify the types of job that are most at risk from the spread of robots and AI, and those that are relatively immune – and to explain why. And it discusses which areas might see more jobs created and even speculates about what sorts of new jobs might appear.

The discussion begins with an analysis of the scope for driverless vehicles, which leads on to an analysis of the scope for replacing humans with robots and AI in the military. It then goes on to discuss the position of various sectors supposedly under threat – manual work in general and domestic help in particular, as well as various sorts of routine mental work.

Then the discussion moves on to the expansion of existing jobs and the creation of new ones, featuring increased employment in the leisure industry and various activities where "the human factor" is all important.

Self-driving vehicles

One of the most widely talked about categories of jobs supposedly at risk from robots and AI is drivers. It is worth spending a fair bit of time on this example because it highlights both the potential and the problems with AI, with clear implications for other sorts of employment.

The number of jobs at risk from the advent of self-driving vehicles is potentially huge: bus drivers, truck drivers, taxi drivers, chauffeurs, delivery drivers, and many more. A 2017 trucking industry report predicted that by 2030, out of 6.4 million trucking jobs in America and Europe, about 4.4 million of them could have disappeared as "robots" do the driving. "Truck delivery driver" is the most common occupation in nine US states.[3] (Mind you, the effects have not yet been felt. Indeed, in 2018 America, trucking jobs are in massive excess demand as the strong economy and the boom in internet shopping combine to create a surge in long-haul shipping. In response, pay rates for truckers have been soaring.)

The implication of the widespread adoption of driverless vehicles would be a major reduction in costs across the economy, especially a large geographically spread-out one such as the USA. According to Alec Ross, drivers account for 25–35 percent of the costs of a trucking operation.[4]

But things aren't quite as straightforward as they seem. On driverless vehicles there is a yawning gap between the hype and the reality. I will start by discussing the positive case and reviewing the potential impact, before moving on to the criticisms and problems.

The potential and the promise

Driverless cars are not a fantasy: they are already working – admittedly only, so far, in restricted areas, such as parts of the city of Phoenix, Arizona. The state of California has recently approved new rules allowing driverless cars to operate without a human driver sitting behind the wheel. In the UK the Chancellor of the Exchequer, Philip Hammond, told the BBC that he aimed to have "fully driverless cars" in use by 2021.

About 50 companies, including Alphabet, Apple, Ford, GM, Toyota, and Uber, are already testing self-driving cars in California. Indeed, more than a hundred trials of autonomous vehicles are currently taking place around the world. Moreover, according to the companies developing them, the performance of self-driven cars is already impressive and is improving all the time. All these companies have invested huge sums and clearly believe that driverless vehicles are the future. But, of course, this does not necessarily mean that they are right – or that they will get a good return on their money. I will shortly have provided you with enough evidence, I hope, to enable you to make up your own mind on this question.

You can readily understand the reasons for the enthusiasm for driverless cars. Partly, this is the matter of cost saving, referred to above. But the advantages go well beyond this. Human drivers kill 1.2 million people a year, and additionally injure between 20 and 50 million people. Some estimates put the cost to middle income countries at about 2 percent of annual GDP. And these accidents are generally caused by the common failings of human beings – drunkenness, tiredness, sickness, and distraction.[5]

Moreover, imagine all those old and infirm members of society who can no longer drive, as well as all those who never could drive, who, in a world with driverless cars, would not need to drive. They would have as much mobility as the rest of the population, freed from the inadequacies of public transport and the expense of taxis. Meanwhile, parents would be freed from the regular chore of dropping off and picking children up from parties, ballet lessons, football games, and so forth. And going to the pub or to a party would no longer bring on the agonized choice over whether to drive and not drink, or instead to fork out for a taxi.

Over and above all this, there is the saving of the time that we all spend driving ourselves to work, to meet friends and family, doing the shopping, going on holiday, or performing some errand or other. It is fine if we actually enjoy the driving but most of us don't – particularly not congested urban driving or being stuck in a motorway traffic jam. Think how much better things could be if someone – or rather something – else did the driving. We could watch films, learn a language, work, drink to our heart's content, or go to sleep. What bliss! Although the effects would not show up in the GDP figures, the result would surely be an increase in human wellbeing.

Wider still and wider

Furthermore, if driverless vehicles really take off, there are potentially huge implications beyond the redundancy of human drivers. The ultra-enthusiasts talk of a transformation of urban land use as people *en masse* forgo their individual cars and are transported in driverless, shared-use, electric vehicles. It is quite possible that car ownership would fall sharply as people predominantly chose to take rides in driverless vehicles from a floating pool.

A joint study by the World Economic Forum and the Boston Consulting Group sees substantial scope for the sharing of rides in driverless cars, thereby undermining the market for public transport.[6] Elon Musk has said that "owning a human-driven vehicle will be similar to owning a horse – rare and optional.[7]

The results would include fewer cars needing to be built (as well as sold, repaired, insured, etc.). Additionally, there would be less demand for space to park cars that remain idle most of the time. While they are waiting for users, driverless cars can be parked end to end and stacked. This could potentially transform urban landscapes and free up much scarce space for other uses. In 2016, Google's Chris Urmson told a US Congressional Committee that in the US parking takes up an area the size of Connecticut. By implication, if everything went according to plan with driverless cars, this space could be freed up for other uses.

And the potential implications go wider still. Perhaps traffic wardens would also disappear as the need to restrict parking becomes less of an issue. In any case, if and when vehicle parking does still have to be regulated, presumably a nice robot would do the job of traffic warden, slapping a parking fine on a driverless vehicle, to be picked up by – well, there's a question that we will return to later. I am sure that the loss of your local friendly traffic warden will be much lamented – even if he or she is replaced by a robot.

Interestingly, if this happens, this will be an example of a job that will have come and gone in pretty short order. Fifty years ago, the job of traffic warden didn't exist. If the enthusiasts for driverless cars are right, in less even than a normal human life span, it could now be headed for the scrap heap.

And there are potentially major effects on the insurance industry. In the USA, vehicle insurance accounts for about 30 percent of all insurance premiums. There are particular issues with regard to who bears the liability when a driverless vehicle is involved in an accident. Doubtless this would provide a fruitful area of business for insurance companies – as well as, inevitably, for an army of lawyers. But a sharp reduction in the number of vehicles to be insured would deal a heavy blow to insurance companies' revenue streams.

Serious problems

So, the implications of driverless vehicles are potentially huge. But it is now time to take account of the more skeptical view about their prospects, before making an overall assessment.

The idea of driverless vehicles has been doing the rounds for almost as long as vehicles have existed. General Motors introduced the idea at the 1939 World's Fair in New York. Of course, since then the technology has become much more capable than what could even be imagined then. (The 1939 concept was of a radio-guided car.)

Indeed, this whole issue has been characterized by overoptimism since then – and it still is. In 2012 Sergey Brin, the founder of Google, said that driverless cars would be available to Google's employees within a year and would be available on the commercial market in "no more than six years." That means 2018. At the time of writing, we are still waiting.

Sergey Brin is far from being alone in his overoptimism. In 2015 Mark Fields, the CEO of Ford, said that fully autonomous cars would be on the market by 2020. Well, as I write this in early 2019, I suppose he *could* still be proved right. But this does not seem at all likely.

In practice, there is no reason for the three elements of the ultra-radical vision of the future of car travel – driverless, shared-use, and electric – to occur together. We need to unpick this attractive, and supposedly inevitable, triad. Shared use runs against a deep-seated desire for privacy, individuality, and (at least some semblance of) control. Meanwhile, the widespread use of electric cars comes up against the shortage of battery and charging capacity. So, these two parts of the triad that is supposedly destined to transform our society are very far from being nailed on. More importantly, we could see the widespread use of shared-use vehicles or electric vehicles, or both, without seeing a large-scale move to driverless vehicles.

For, even without the ride sharing and the switch from petrol to electric, the widespread use of driverless cars is not as straightforward as is usually implied. Feasibility is not the issue. Safety is. Demis Hassabis, one of the founders of DeepMind, said in May 2018: "How do you ensure, mathematically, that systems are safe and will only do what we think they are going to do when they are out in the wild."[8]

His misgivings are fully justified. Despite the claims of the manufacturers and developers of driverless vehicles that they are ultrasafe, a 2015 study from the University of Michigan discovered that the crash rate is higher for driverless vehicles.[9] The study suggested that, when they occur, crashes are almost always not the fault of the driverless cars. The problem seems to be that human drivers find it difficult to interact with other vehicles when the latter are driverless. This is such a problem that some tech companies are trying to make driverless cars less robotic, even inducing them to cut corners, be aggressive, and inch forward at junctions.

In fact, things aren't quite so simple as even this might seem to imply. For all the boasts about what their autonomous vehicles can do and the reports that their vehicles have passed so many tests with flying colors, the claims of the manufacturers and developers of autonomous vehicles cannot be taken seriously. For these tests are usually conducted in secret and without independent verification. We do not know – and we are not allowed to know – the road and weather conditions that the vehicles were subjected to, nor how far they were dependent on any human intervention.

It is significant that so much of the experience with driverless vehicles so far has been in locations like Phoenix, Arizona, a place blessed with a predictable and attractive climate and good driving conditions. No snow, no fog, and no convoluted road systems or random congestion. A more serious test would be to put these cars through their paces in London, Moscow, or Istanbul – in February.

The degree of human intervention

Legislators, courts, and insurance companies are having to deal with some very tricky issues created by driverless vehicles. Under new UK legislation, drivers of self-driving cars must not take their hands off the wheel for more than a minute. And in April 2018 a motorist was banned from driving after being caught on the M1 motorway in the passenger seat, with the driving seat vacant as the AI "drove" the car.[10] The British government is planning to scrap the requirement for a "safety driver" to enable advanced trials on public roads of fully automated vehicles by the end of 2019. It will be interesting to see how far this goes.

The American experience invites skepticism. In June 2018 a report from the Tempe, Arizona police department into the crash of a "self-driving" Uber vehicle that killed a 49-year-old woman who was crossing the street said that the crash was "entirely avoidable." The report said that evidence showed that "the safety driver" had been distracted as she had been streaming a television show on her phone. She could face charges of "vehicular manslaughter."

A 2018 report from the Association of British Insurers warned that car manufacturers and drivers needed to make a distinction between "assisted" driving, where the car's computer helps the driver with selected tasks, or indeed performs them, and "automated" driving in which the computer effectively takes control. According to current law, the motorist is required to take control in problematic circumstances. But that means that the driver must keep their eyes on the road and must stay constantly alert.

In fact, among the developers of driverless vehicles and AI enthusiasts, such distinctions have gone much further. Discussion is dominated by the so-called six levels of autonomy, ranging from level 0 to level 5. At level 0 there is no autonomy at all: human drivers do everything and there are no intelligent system aids. At level 1 the driver must control all normal driving tasks, but limited tasks can be assigned to an intelligent system installed in the car, for example parking assistance. Level 2 is characterized as "semi-automation." The driver must monitor the system and the environment continuously. But the driver can assign control of the vehicle, including steering, acceleration, and braking, when conditions are appropriate.

At level 3, described as "conditional automation," the car drives itself, but the human driver needs to be ready to resume control when necessary. And the system knows its own limitations and will periodically request driver assistance.

At level 4, described as "high automation," cars drive themselves without any input from the driver, but "drivers" retain control over pedals and the steering wheel in order to be able to take over when necessary. At level 5, no human involvement at all is possible, even in emergency conditions.

Levels 2 and 3 have proved to be unsafe as humans at the wheel become inattentive and distracted and are therefore unable to intervene promptly when necessary. There is a well-known and tragic case of an accident in Florida in 2016. Joshua Brown, a keen advocate of Tesla cars, had put his car on autopilot but the car's sensors failed to register that a large truck was crossing the car's path. The car steered itself under the truck, killing Brown.

Recognizing these problems, within the industry the great ambition now is to develop vehicles at level 4. Actually, a level 4 autonomous truck is awaiting regulatory approval in Sweden. There is no driver cabin or controls but, if need be, the vehicle can be operated remotely by a supervisor sitting hundreds of miles away, who can supervise up to 10 vehicles at a time. This vehicle, known as a T-Pod, would initially travel only six miles a day and would be allowed to operate on only 100 miles of public roads where it might encounter vehicles with human drivers.[11] Yet substantially the same issues arise, unless the car really is capable of dealing with all circumstances. If not, how can you be ready to intervene in an emergency if you are not constantly alert? What if you are drunk – or asleep?

And what is the point of going driverless if you, the human driver, the "safety driver," or whatever they call you, have to pay attention the whole time? Isn't the point of going driverless that you, the erstwhile driver, can read the newspaper, fall asleep, or get drunk?

A further problem derives from the deskilling of drivers as a result of relying on technology. This is ironic because it is precisely when, for whatever reason, the technology fails, or cannot cope with a particular set of circumstances, that intervention by humans is required, humans who are supposed, at that moment, to be more capable than the machines/automatic systems that have failed. But how can they be more capable if they have been used to sitting passively while an automatic system did all the work and made all the decisions?

This phenomenon is not restricted to driverless cars. The same thing applies to aircraft and ships. The most significant, and tragic, example of this is the loss of the Air France flight A330 which plunged into the Atlantic in June 2009, killing all on board. It seems that the crew could not handle the plane properly when the systems malfunctioned.

This isn't an issue with level 5 automated driving because in this case human intervention isn't even possible. But, to reach level 5, driverless cars will need to be able to cope with all weather conditions, including fog, blizzards, and snow, be able to distinguish between a football being kicked into a road and the child chasing it, distinguish between a dog and a child, and be able to negotiate their way along streets crowded with people, often doing unpredictable, and sometimes apparently nonsensical, things. None of these things can driverless cars readily do now.

Moreover, a way must be found of coping with unmapped roads and changes to road layouts. Wonderful though GPS is, it has been known to lead people up the garden path, literally and metaphorically, and even, in some cases, into the sea.

Robert Dingess of Mercer Strategic Alliance, which is a lobbying firm that specializes in automobile technology, has put the current state of play most pithily. He notes that the manufacturers have become very good at "developing self-driving systems that operate safely 90 percent of the time, but consumers are not happy with a car that only crashes 10 percent of the time."[12]

As so often with evangelists for new technology, some blame humans for the shortcomings of their baby. With regard to driverless vehicles, Carlos Ghosn, the former CEO of Nissan, well before his fall from grace, said: "One of the biggest problems is people with bicycles. The car is confused by [cyclists] because from time to time they behave like pedestrians and from time to time they behave like cars."[13] Just imagine: those pesky cyclists interfering with the successful operation of this wonderful new technology! Surely, steps will have to be taken to ensure that they don't.

The pilot comparison

According to AI enthusiasts, another job that is clearly destined for the scrap heap is airline pilot. But people have been suggesting this for ages. As is well known, planes are already mostly flown by computers, with pilots controlling the plane for only a fraction of the flight. Is this combination of humans and computers superior to having the plane flown by humans alone? Probably. A high proportion of aircraft crashes are caused by human error.

Equally, having a plane flown by both a pilot and a computer is probably superior to having the plane flown only by computer – partly because of the incidence of extreme scenarios when a human pilot would perform better, or computer systems malfunction. There is also the issue of public perception. This ensures that there is no realistic prospect of passenger aircraft being flown "automatically" without a pilot. I really cannot imagine 200-plus people willingly boarding a metal or fiberglass tube to be propelled through the air at massive speed and high altitude, controlled only by a computer, even if this is now called a robot or an AI.

In order willingly to subject themselves to this, surely the passengers will themselves have to be robots. (Perhaps they will be.) Whatever the accident statistics say, human passengers will always worry about the unforeseen event that a human pilot could and would be able to deal with and a machine could not. Fans of the actor Tom Hanks will surely readily recall the film *Sully: Miracle on the Hudson*, and nod sagely in agreement.

This has a close bearing on the vexed question of how far driverless vehicles are allowed to proceed. For airline pilots, substitute coach driver. Will people really be prepared, *en masse*, to take long journeys in a coach completely in the hands of AI? And put their children on such a coach? I doubt it. It is tempting to believe that at least commercial vehicles and lorries are categories of vehicle that escape this problem. But I am not so sure. The destructive power of an ordinary commercial vehicle, never mind a heavy-duty lorry, running amok is quite terrifying.

Similar points apply to Amazon's idea of delivery by drone. Of course, it is technically feasible. That is not the point. The point is the enormous capacity for accidents involving a five-pound payload suspended hundreds of feet in the air.

The importance of human judgment

There is a chilling story from the Cold War that illustrates the importance of the human factor. At a time when tensions were already high because of the shooting down by the Soviet Union of a Korean aircraft, killing 269 passengers, a Soviet early-warning system reported the launch by the USA of five missiles at the Soviet Union. The Soviet officer responsible for this early-warning system, Stanislav Petrov, had minutes to decide what to do. The protocol required that he should report this as a nuclear attack. Instead, he relied on his gut instinct. If America was really launching a nuclear attack, he reasoned, why had it sent only five missiles?

So, he decided that it was a false alarm and took no action. He was right. It turned out that a Soviet satellite had misread the sun's reflections off cloud tops for flares from rocket engines. It is now widely acknowledged that Petrov's judgment saved the world from nuclear catastrophe.

Despite involving the saving of the world from a real disaster, this true story has a bitter ending. Stanislav Petrov was sacked for disobeying orders and lived the rest of his life in drab obscurity. (That was the Soviet Union for you.)

Could we imagine entrusting the sort of judgment that Petrov made to some form of AI? This conjures up the scenario depicted in the now very old but still rewarding film *Dr. Strangelove*, starring Peter Sellers, in which, once the USA has embarked on a nuclear attack, it is impossible to stop it from going ahead. The system takes over and will not respond to attempted human intervention.

Further problems

Safety concerns and issues about legal liability pose serious obstacles. At the very least, they are likely to mean that the actual deployment of driverless vehicles for both commercial purposes and "leisure" travel will lag well behind their technical feasibility.

But, quite apart from the important, but narrow, matter of driver, passenger, and third-party safety, there are three other problems that driverless cars throw up. Of course, accidents are a serious source of concern, but suppose that a nefarious person or organization was able to hack into the system controlling a vehicle or set of vehicles? This risk is scary enough with regard to disaffected individuals or criminals, but think about terrorist groups. If there were no human drivers in vehicles, including planes and cars, and a terrorist group were able to hack into the computer systems controlling these vehicles then they could deliberately cause mass slaughter on an industrial scale by turning the whole transport system into a weapon.

Second, there are also serious issues about privacy because driverless cars, of course, carry cameras that look outside and inside the car and transmit data about what they see. Who will own this data and who will control its transmission and availability?

It falls appropriately to an economist to raise a third key issue that seldom seems to be discussed by the techies, namely cost. The kit necessary to enable a car to operate autonomously is phenomenally expensive. Just because something is technically feasible does not make it necessarily economically desirable, as the operators of the Concorde airplane painfully discovered.

If driverless taxi services or commercial vehicles continue to need a "safety driver," then there will be scarcely any cost saving from going driverless over using a conventional driver. But, worse than this, driverless systems demand an army of software engineers to solve difficult road problems, such as a blocked lane. And these software engineers are typically extremely expensive. Accordingly, far from saving costs, it is more likely that going driverless will increase them.

Advocates of the vision of driverless vehicles sometimes try to counter the point about "safety drivers" still being needed by pointing out that these "safety drivers" can still provide some of the ancillary services provided by drivers now, such as helping passengers with their bags, helping them in and out of the vehicle, and chatting to them during the journey. This is true, but they cannot do this anymore when they are not actually driving the vehicle than when they are. And while they are there in the car, they cost just the same. So, what's the point?

Overall assessment

Elon Musk, the boss of Tesla, has warned against setting the safety requirements of driverless vehicles too high. After all, he reasons, since human error when driving is responsible for a large number of fatalities, there is room for driverless vehicles to cause some fatal accidents that a human driver could have avoided and yet for the introduction of driverless cars still to reduce the overall accident rate and the number of fatalities. But I don't think this line will wash with either regulators or the public.

Accordingly, the scale of the replacement of human drivers by AI is likely to be much less than the enthusiasts allege. No level 3 automated car is being sold in the market today. Currently, the technology is somewhere between levels 2 and 3. Just to remind you, these are the levels that require substantial human intervention, thereby diminishing the advantages of going driverless and/or running huge safety risks.

This is not to say, though, that there is no scope for driverless vehicles. Already, cars can drive themselves, unassisted, on motorways and can park themselves. These features can and do bring some benefits to their users. And doubtless over time more people will want to avail themselves of these services. But, of course, this is a long way from the human user, the erstwhile "driver," being able to switch off altogether. And while human drivers are still needed to be capable of taking control then most of the much-hyped economic consequences of driverless vehicles simply won't happen.

Mind you, journeys with restricted routes, where the scope for things to go wrong is limited, will surely be wide open to the full replacement of humans by AI drivers. In fact, we already experience such AI drivers quite frequently. Driverless rail shuttles at airports and underground trains have been common for some time.

And there is surely scope for more widespread replacement of humans on railways. In Scotland, Glasgow Subway announced in

September 2018 that its new trains would be driverless and totally unstaffed by 2021. (The proposal, though, ran into strong opposition from the transport unions. Being able to escape from the strikes and obstructionism of the transport unions would be one of the major advantages from driverless trains.)

And, for road transport too, there is surely scope for the replacement of humans on restricted routes such as dropping off and picking up passengers at different car parks at airports, or even perhaps driving along urban tramways or even restricted urban bus routes. In fact, four-person driverless pods to shuttle passengers between Heathrow's Terminal 5 and the car park have been in operation since 2011. And there is clearly scope for an increase in the use of driverless tractors and other farm vehicles operating on agricultural land, away from clusters of people.

Norway has introduced a driverless ferry. Mind you, so far it has been confined to a 320-foot stretch of water in Trondheim. The journey takes only 60 seconds. I am sure that there will be plenty of instances of such "auto-ferries" moving people and goods short distances across restricted waterways. But this is a far cry from a fully automated ocean-going cargo ship or cruise liner sailing without a captain.

Whether the vehicle is a car, a plane, or a boat, what has been achieved so far, and what is likely to be achieved in the foreseeable future, is a far cry from the wholesale replacement of humans by AI drivers that the AI enthusiasts think is imminent. In that case, the benefits of driverless vehicles have been massively oversold. Indeed, in the UK, a recent House of Lords Committee on the subject came to a skeptical conclusion and criticized the UK government for falling for the hype surrounding driverless cars. It recommended that research should be concentrated elsewhere, including marine and agriculture.

There is some evidence that in private this is increasingly recognized within the industry, although people are afraid to speak out for fear of the consequences. The whole thing has an aura of the emperor's new clothes about it. It reminds me of the excessive enthusiasm about all things digital in the run-up to the bursting of the dot-com bubble.[14] Of course, there was much to admire about some of the ideas and the companies that proliferated then. And some have more than endured: they have grown like Topsy and transformed the business landscape. But there was also an awful lot of dross that was blown away as soon as the bubble burst and people recovered their senses.

If it is right to think of the time and treasure spent on driverless cars as reflecting a bubble, then there is going to be a serious reckoning when it bursts. The new tech monopolies have surplus billions to burn, but the car manufacturers don't. Coming on top of all the other challenges that the car industry faces, for those companies which have invested heavily in it, the dying of the driverless vehicle dream could pose an existential crisis.

Military applications

Closely related to the issue of driverless vehicles is the possible use of robots and AI in military applications. The key issue is the same: do robots and AI hold out the prospect of replacing humans – in this case, soldiers, sailors, and airmen/airwomen – with machines?

The application of robotics to military operations is nothing new. Robots were used during the Second World War and the Cold War. In the Second World War, both the Germans and the Soviets used automated vehicles to carry large amounts of high explosives. These were essentially mobile robots, relying on human guidance and were essentially teleoperated or remote-controlled.[15] However, their military value was limited, given their high cost and low speed.

But things have moved on a lot since then. Currently, the majority of military robots are teleoperated and used for locating enemies, surveillance, sniper detection, and neutralizing explosive devices. Unmanned aerial vehicles (drones) are currently the most common use of military robotics and, although these are sometimes used for surveillance purposes, they are often armed with missiles. The inventory of unmanned aircraft in the USA grew from about 50 in 2001 to over 7,500 by 2012, of which about 5 percent were armed.[16] In 2007 Congress announced a "preference for unmanned systems in acquisition programs for new systems."

Because robots are unaffected by emotion, adrenaline, and stress, some analysts believe that replacing humans with autonomous robots could lead to fewer atrocities and war crimes. It can be argued that, because of greater accuracy and the absence of all the usual human defects, autonomous AI systems of warfare will "not only reduce the number of humans put in harm's way but can also significantly add to the security of those who must face danger."[17]

So, given the advantages of military robots, does this imply a reduced demand for military personnel? H. R. Bart Everett, a Navy robotics pioneer, does not see robots replacing humans but rather envisages a "human-robot team, much along the lines of a police dog and its handler."[18]

As in so many other cases, it seems that in military applications the ability of robots and AI to economize on human inputs has been overhyped. The US Air Force Vice Chief of Staff, General Philip M. Breedlove, works with drones. He recently made a remarkable statement: "Our number one manning problem in the air force is manning our unmanned platforms." He was referring to the human workers needed to repair and maintain drones and analyze the videos and surveillance information obtained by drone sorties. Apparently, the US Air Force reckons that to keep an unmanned Predator drone in the air for 24 hours requires 168 workers. A larger drone may need 300 people.[19]

So, the upshot may well be a reduction in the number of frontline military personnel trained and primed to face the enemy, and an increase in the number of personnel tending to all the various needs of military robots and AI. This job shift would be fully in accordance with experience in the wider economy.

The manual factor

The assessment of the scope for driverless vehicles and for the replacement of humans by machines in the military provides useful background for an analysis of job prospects in other sectors. In some areas it is a similar case of hype and overselling, but in others there is scope for the replacement of humans by machines to go much further than is likely with cars and other modes of transportation. And, as in the military, there is great scope for the *type* of job available to humans within any given sector to change radically.

It is widely believed that the challenge from robots and AI will be felt most acutely in manual occupations. Actually, it is not true that all manual jobs are acutely under threat. As noted above, robots continue to be poor at manual dexterity. According to the cognitive scientist Steven Pinker: "The main lesson of thirty-five years of AI research is that the hard problems are easy and the easy problems are hard."[20] Accordingly, many skilled, manual jobs look safe for the foreseeable future. These include plumbers, electricians, gardeners, builders, and decorators.

That said, research from Mace suggests that by 2040, out of the 2.2 million jobs in the construction industry, 600,000 could be automated. According to recent reports, there could eventually be significant job losses in the ranks of those who dig up roads to identify and then repair faults in pipes used for water, gas, and sewage. The British government has invested over £25 million in a project researching the development of mini-robots which would travel down the pipes to identify and fix the problems without roads having to be dug up. This would be rather like keyhole surgery for the ground. And, just as with keyhole surgery on the human body, the robots in the pipes that repaired the faults would be controlled by humans above the surface.[21]

At the low end of the job sophistication scale lies a sector where is it thought that jobs will be largely eliminated – retail. Recently, a new "Amazon Go" store opened in Seattle, designed to have no employees serving in it at all – and no queues at checkout. Banks of cameras and AI equipment monitor what customers take from the shelves as they saunter about the shop and payment is automatically deducted from their account when they leave. A report from Cornerstone Capital published in 2017 estimated that these technologies could eliminate about 7.5 million retail jobs in the USA, including 3.5 million cashiers.

But it has to be said, so far the evidence of retail outlets with minimal human staff is not altogether convincing. Britain's first "shopbot," known as "Fabio," was sacked after a week for confusing shoppers in the Edinburgh flagship shop of Margiotta, the Scottish supermarket chain, where it was being trialed. When asked by customers to say where to find beer, Fabio replied: "It's in the alcohol section."[22] Of course, this answer was factually correct. But it wasn't exactly very helpful.

Retailing is an interesting example of an activity where not only are traditional jobs being lost but the nature of employment is being transformed. The job of check-out assistant is on the way out and the loss of this boring, repetitive job will surely not be much lamented – provided those who used to do this job find decent employment elsewhere. Meanwhile, there is scope for a different sort of employment in shops as assistants help, guide and advise customers about what to buy, what goes well with what, and about the different characteristics of the various options. This new sort of retail job involves the human element and requires more skill and knowledge, while the boring, repetitive stuff is left to machines.

Further up the supply chain, Ocado, the British internet supermarket, is reinventing itself as a supplier of warehouse services to retailers around the world. It has recently landed a contract with the Canadian grocer Sobeys which will launch an online shop with orders picked from an automated warehouse in the Toronto area. There won't be many jobs for humans in the warehouse. Battery-powered robots are at the center of Ocado's model.

Mind you, robots are still not very good at picking things from the shelves. As a result, warehouses are now using "the Jennifer Unit," which is a headset that tells human workers what to do. The economist Tim Harford envisages this as a possible harbinger of bigger things to come. As he says: "If Robots beat humans at thinking, but humans beat robots at picking things off shelves, why not control a human body with a robot brain?"[23]

In restaurants, I doubt that robots will ever take over the role of waiting at table, although they might at some point be able to help with the washing up. Interacting with the waiter is part of the restaurant experience. Interacting with a robot just won't be the same.

Domestic help

Another example of a manual job type that seems to be under serious threat, but turns out not to be, is housekeeping and the provision of various domestic services. iRobot's Roomba can vacuum the floor but that is about all. Although more than 10 million of them have been sold they won't be straightening the magazines on the coffee table, folding towels, or plumping the cushions. They cannot accomplish any of these simple tasks.

This is particularly important because the increase in income made possible by the advances of AI will potentially unleash an enormous demand for domestic helpers of all sorts. On all the current evidence, these domestic helpers will not be robots but humans.

This prospect readily conjures up a bygone era. Before the First World War, it was common for the majority of the middle class, stretching down even into the ranks of clerks and administrative assistants, to employ at least one domestic servant. Higher up the income (and status) scale, people employed umpteen servants – maids, butlers, cooks, gardeners, cleaners, footmen, and heaven knows who else. Among the really grand there was even a degree of competition over the numbers and sartorial splendor of the more visible of these retainers.

It is interesting to reflect on the economic reasons for the near demise of domestic service. There were several factors. The erosion of middle- and upper-class incomes from taxation played a part, along with a change in social attitudes. Equally, in the 1950s, 60s, and 70s, the emergence of many labor-saving devices in the home, such as washing machines, vacuum cleaners, and dishwashers, reduced the time needed to be spent in keeping a household going.

But a major reason was surely the huge increase in the productivity of labor in factories made possible by mechanization. This increased the real wage that workers could earn in employment outside the home, which increased the wages and salaries that employers would have to pay for domestic staff in order to be able to compete. If large numbers of jobs are now to be lost to robots and AI in manufacturing and the office sector, this fundamental force that has been operating against the employment of domestic staff could go into reverse.

Domestic employment could potentially be a major source of future jobs, even if the "upstairs downstairs" world is gone for good. Of course, many people already employ some form of domestic help, especially cleaners and nannies, albeit part-time and usually resident outside the home. This could go much further. Prosperous middle-class people who, in today's world, are notably short of time rather than money could readily employ at least one domestic helper (augmented by some robots), probably part-time and not living in, to do a series of domestic tasks, including shopping and driving to pick up and drop off family members. Meanwhile, those who sit at the top of the income pile could readily compete for status over the number and magnificence of their domestic retinue, as they used to in the old days.

Routine mental jobs

Many of the jobs that are most at risk from robots and AI are not ones that are usually described as manual. At the low end, check-in assistants at airports are on their way to being checked out. At the other extreme, many of the jobs in the fund management industry can be handed over to AI applications. In 2017 the mammoth fund management firm BlackRock sacked seven fund managers and transferred billions of dollars which they used to manage to an internal unit called Systematic Active Equities, a computer-powered quantitative investment unit.

And in 2018 it emerged that the German bank Commerzbank is experimenting with AI technology that can generate analysts' research notes on companies and sectors, thereby allowing it to dispense with umpteen human analysts who have traditionally been among the most highly paid employees.[24] (I must say, I was under the impression that investment bank analysts' reports were already written by some form of AI.)

And property valuers are similarly at risk of being replaced by AI. Apparently, it can perform the task more quickly, cheaply, and accurately. And this is pretty significant. In the USA the property valuation business is worth about $12 billion per annum and employs about 80,000 property appraisers.

Similarly, routine legal work can now be done by AI. Mind you, these trends do not necessarily mean that there will be fewer lawyers (unfortunately). It seems that the use of AI applications to do routine legal work has so reduced costs that a whole class of legal projects that would previously have been uneconomic are now feasible. And it turns out that junior lawyers are needed to set up the initial stages and get the AI apps working in the right way.[25]

Another example of a vulnerable job that is not manual at all is translating and interpreting. At the beginning, digital translation services were a joke. Actually, to a large extent they still are. But the rate of improvement has been immense, even though they do not function at a level that many people find acceptable. Soon, though, at the click of a mouse, it should be possible to translate anything to any language at a very high level of competence, indeed beyond the capability of most human translators.

In March 2014, Skype introduced real-time machine translation. In June 2013, Hugo Barra, Google's top Android executive, said that within several years he expects a workable "universal translator" that could be used either in person or over the phone.[26]

Mind you, even if these improvements continue and routine translation work is all done by machine, there will still be people who will make their careers out of being language experts. But these will be at a very high level, and much of their work will concern the overseeing and improvement of AI-driven translation and interpretation services. Of necessity, there will be far fewer of these senior language experts than are engaged in translation and language services at present.

One activity that you might think of as bound to be the preserve of humans is interviewing candidates for jobs. In fact, though, at leading edge firms, especially in finance, much of this activity is now performed by some type of AI. For decades, candidates for employment at big firms have had to undergo automated online tests, with the initial weeding-out done without human intervention. Now, though, interviews are increasingly being carried out by computers. Predictably, job candidates are reacting by using AI to prepare themselves to get through job interviews conducted by other forms of AI. A fintech start-up called Finito offers to coach job candidates with its own AI.

In China AI is helping to censor the internet, thereby reducing the rapidly increasing burden on human censors. Apparently, machines are very good at recognizing sexual content, and so the human censors need to spend very little time on that, thereby freeing up capacity to concentrate on other, more important, things, such as censoring people saying nasty things about the Communist Party, or President Xi Jinping.

Job expansion

So much for job destruction. What about job creation? According to a joint study by the World Economic Forum (WEF) and the Boston Consultancy Group (BCG), by the year 2026 12.4 million new jobs will be created in the USA.[27] The widespread use of robots and AI applications is itself going to lead to new jobs – designing and making robots, developing their software and AI applications, teaching humans how to get the most from their AI helpers, perhaps even counselling people who have serious problems with their AI relationships.

There are also going to be umpteen jobs policing the boundaries between humans and robots, dealing with legal and regulatory issues and perhaps even considering and monitoring whatever ethical issues are thrown up by AI, perhaps especially in relation to the use of big data. (I discuss these issues in Chapter 7.)

But the sector predicted by the WEF and BCG study to enjoy the largest net job creation is one already very much in existence: healthcare. It will lose only 10,000 positions while 2.3 million positions will be created.

This should not be surprising. The essence of the coming technological change is to remove umpteen mechanistic jobs from human beings and to leave to them the realm of the truly human. This is, after all, the area where human beings will most obviously have a comparative advantage. And this field will be enormous. Personal care, particularly for the aged, is a major case in point.

We are constantly being told that in all Western societies care for the aged is inadequate. Resources are in short supply. There simply aren't enough caregivers to go round. Yet, with the number of older members of society set to rise enormously, demand for caregivers will expand sharply.

Well, in future there will be far more caregivers. I am not suggesting that the caregivers will be robots. After all, who would want a home visit from a robot? And who would entrust the care of their elderly relative to one? (Come to think of it, though, I can think of one or two.) The extra caregivers will be humans, released from other, *non-caring*, jobs in the economy.

But here again, robots and AI can make a contribution – not by replacing human caregivers but by helping them. I can readily imagine a caregiver visiting an elderly person's house accompanied by a range of machines to help with the job. The caregiver could set one machine to do the cleaning and another to start washing the client's hair, while the caregiver talks to the client about their week and what else they might want done.

Accordingly, the use of robots and various AI devices will enable a radical increase in the productivity of caregivers. As with other labor-saving developments since the Industrial Revolution, the number of caregivers employed will not go down because as people get richer (and older), the demand for these services will rise considerably.

Actually, such robot helpers exist today, but they are not very effective. There are some machines that can lift and move elderly people. The trouble is that they are extremely heavy, weighing up to 10 times a person's weight. And they are very expensive.[28] Doubtless, they will become lighter and cheaper.

Enhancing existing jobs

There are many jobs where, although it is often reported that robots are set to replace humans, in fact they are destined to assist them, and in the process improve the quality of what they do. As we have seen, that is true of care for the elderly.

Surgery provides another medical example. Supposedly, robots are increasingly taking on the job of surgeons. London's University College Hospital is home to two da Vinci Xi robots, created by Intuitive Surgical, which is based in California. In 2017 they removed 700 prostates and bladders.

But when we say "they," this is not quite accurate. The robots wield the scalpel all right, but they are controlled by a surgeon who sits, a few feet away, looking at a 3D screen.[29] So the robot is actually a very sophisticated tool. On the face of it, the results are dramatic: much less invasive surgery, more accuracy, less risk of mishap, fewer complications, and a better recovery profile.

Mind you, it is not all sweetness and light. First, there is the cost. In the USA in 2013 some 1,200 surgical robots were sold at an average cost of $1.5 million each – not exactly a mere bagatelle. Second, there is some concern about the safety record. According to the *Journal for Healthcare Quality*, there have been 174 injuries and 71 deaths related to da Vinci surgical intervention.[30] Of course, as so often, it is impossible to know whether any of these deaths or injuries are precisely because of the employment of a robot surgeon or whether they would have occurred anyway. And whatever the safety record so far, I am sure that it can only improve.

There have been other dramatic advances in the medical field. It is alleged that to keep up with published medical research a human doctor would need to read for 160 hours a week, which is clearly not feasible. But AI has no such restriction.

It is perhaps unsurprising, therefore, that AI also has an advantage in diagnosis. Google's DeepMind has developed AI to diagnose diseases by analyzing medical images. According to Samuel Nussbaum of the private healthcare company Wellpoint, *Watson* has a diagnostic accuracy rate for lung cancer of 90 percent, compared with 50 percent for human physicians.[31]

Developments in medical diagnosis are likely to lead to more frequent tests and earlier diagnoses. Soon an attachment to our smartphones will deliver instant test results on blood glucose, blood pressure, voice, breath, and so on, and will deliver instant initial assessments of our medical state. This will not lead to less need for the services of medical professionals. Quite the contrary: it is likely to lead to more consultations with medical professionals and more treatment of some sort.

As a result of the use of sensors that track patients' heart rate and blood pressure, thereby facilitating earlier identification of problems and treatment at home rather than in hospital, one possible result is a reduction in the number of people having to spend time in hospital, thereby freeing up resources for critical cases.

In addition, natural language-processing technology enables doctors to transcribe and record meetings with patients with minimal effort and use of doctors' time. A consultant labelling scans at Google's offices said that labelling images for head and neck cancer "is a five or six hour job; usually doctors sit and do it after work."[32] Meanwhile, AI can help with triage in accident and emergency departments and help to reduce "traffic jams" in the flow of patients through different hospital departments.

There is considerable enthusiasm in both medical and official circles as to what AI could contribute to both diagnosis and treatment. In the UK, the then Health Secretary, Jeremy Hunt, said in May 2018 that embracing AI was vital to ensuring the future of Britain's National Health Service (NHS). In the same month, the British Prime Minister Theresa May announced with much fanfare a program to get AI to transform the prevention, early diagnosis, and treatment of heart disease and diseases such as cancer, diabetes, and dementia.

There is no doubting the sincerity of the ambition or the scale of the opportunity. But there are grave dangers here of huge amounts of money being wasted. It has to be doubted whether many of the NHS's practitioners, both clinical and administrative, currently have the skills to make proper use of a huge investment in AI for the NHS. It is noteworthy that the NHS has still not been able to transfer all medical records from paper into digital form after spending billions of pounds on a failed IT project that was supposedly going to do exactly that.

AI users such as the NHS need to make sure that they focus on areas where AI can deliver, even if the improvements are at first marginal, rather than being bamboozled into embarking on major IT projects which may come to nothing.

Employment in the leisure sector

I argued in the last chapter that greater wealth brought about by the increased deployment of robots and AI would lead many humans to choose to work fewer hours and thus to increase their leisure time. Many of the things that people decide to do with their time (although not bridge or voluntary work), will involve the expenditure of money. This leads to what you may think of as a paradox, if not a blatant contradiction. It is a well-known characteristic of working hard that it makes it easier to save money. While you are at work you cannot easily spend money because you simply do not have the time, even if you have the inclination. And when the work finally stops, you are often too tired to think of doing much except resting.

The opposite is true of time not spent working. In order to enjoy this time to the full you need to have money to spend. Here comes the paradox – and the key to many of the employment possibilities that will spring up in the AI economy. Many of the things that you might want to spend your leisure money on involve the employment of other people – especially in the world of sport, entertainment, education, and personal development.

This is even more true now because of a developing shift in consumer preferences. Increasingly, what people want from their leisure time is "experiences." This involves spending more and more money on everything from nights out to holidays and weddings, all of which are highly employment intensive. And it will not be robots that they will want to accompany them, or to share or direct their "experiences," whether these are getting drunk, going on holiday, or getting married. (Mind you, could the role the vicar performs at a wedding be replaced by a robot? Possibly not, although at some of the weddings I have attended it might have been an improvement.)

Increased leisure will give rise to increases in employment opportunities in the conventional providers of hospitality and entertainment services. But over and above this many people will probably try to make more of their leisure. Personal development will become big business, not just personal physical training but also life coaching and spiritual training. Who is going to want to have this done by a robot? Try asking a robot about the meaning of life. Following the example of *The Hitchhiker's Guide to the Galaxy*, it might well answer "42."[33]

This means that as more people choose more leisure, more and more of the jobs on offer will be in catering to the leisure needs of those not working. This will, of course, even include the leisure time of those who work in the "leisure industries." This is not a contradiction. The balance between working in nonleisure activities, working in the "leisure industries," and leisure itself will emerge naturally as the outcome of billions of choices made by people about what to do with their lives. But this does mean that the facilitation of activities to fill people's leisure time will be a major source of employment in the AI economy.

The human factor

This point about the demand for human labor in the entertainment and "experience" industries has broader applicability. Suppose that the cheaper option for all services is to have them provided by some sort of machine. Might you not want to pay a bit extra for the pleasure of having services performed by another human being? Indeed, might this not confer status upon you, the consumer?

In fact, I have come across some people who claim that, other things equal, they would rather deal with a machine than a person, including at supermarket checkouts, airport check-ins, and the like. For such individuals, in competition with machine-provided services, people-provided services would have to come at a discount. But I cannot believe that most people fall into this category.

Suppose, for instance, that the Chinese eventually manage to get their robot waiters to serve you without pouring the soup all over you. Suppose they even manage to get these waiters to serve you with arrogant disdain, or perhaps unctuous politeness, if you prefer. Suppose even that you, the customer, are able to "press a button" to select the service you would prefer, arrogant, unctuous, or umpteen variations in between. Would you not prefer, in fact, to be served by a real live human being? Arrogant, unctuous – or whatever? The same surely goes for a whole range of personal services from home-helps, caregivers, doctors, beauticians, and personal trainers.

Humans enjoy live interaction with other humans. Take entertainment as an example. The world is awash with "distant" entertainment, including sport and music, both recorded and live. Often this is available free, or almost free. And people consume huge amounts of this material. Nevertheless, people also pay serious money to attend live events to see and hear performers in person. This includes all sorts of sporting events, musical performances, and plays. Nor is the attraction of live performance just restricted to the superstars. Plenty of pubs and clubs put on live music and frequently attract a large and enthusiastic local audience.

There is even a reason for choosing to buy some *goods* made by humans – and probably sold by humans. The name that Calum Chace gives for the reason for such preference is "artisanal variation."[34] At the luxury end, this is one of the key reasons why we like antiques. It is striking that an original painting by a great artist such as Monet or Turner commands a hugely greater price than a copy, even if, to the untutored eye, the latter is indistinguishable from the original.

At the everyday end of things, we also often prefer handmade bric-a-brac and individually designed clothes. And, of course, we have a distinct liking for "homemade" food (even if it is made in restaurants, rather than the home) over the preprepared, mass-produced, or microwavable variety.

People watching

The realm of the human will surely include the constant monitoring and mentoring of all sorts of human behaviors. Think of the amount of work and the number of jobs currently emanating from the world of financial investment. Who could have imagined a hundred years ago the number of people pontificating and advising on people's investments, past and future? Writing up yesterday's stock-market report is doing something completely human – that is to say, talking about the actions and opinions of human beings. Yes, in the new world, the bare facts will be recorded and "written-up" by some form of AI, and much of the investing will be done by AI, too. But umpteen jobs will essentially be centered on the thoughts, opinions, and actions of humans. And humans will be the ones to do this work.

Interestingly, this outcome was foreshadowed by Keynes – again. Amusingly, in his *General Theory*, when discussing how stock-market investors decide to buy or sell stocks and shares, he compared much stock-market investment to a beauty contest. He wrote:

"It is not a case of choosing those faces that, to the best of one's judgment, are really the prettiest, nor even those that average opinion genuinely thinks the prettiest. We have reached the third degree where we devote our intelligences to anticipating what average opinion expects the average opinion to be. And there are some, I believe, who practice the fourth, fifth and higher degrees."[35]

He may or may not have been right about investment. (As it happens, I think there is a good deal in what he said.) But he inadvertently put his finger on a likely future source of employment and activity – the fascination of humans with the opinions of other humans about themselves, other humans, and umpteen derivatives therefrom.

Not far from the world of investment is the world of gambling. It has expanded enormously over recent decades – and with it the number of people employed in it, as well as the amount of punters' leisure time taken up with it. As with so much else in the human sphere, it is not just the activity itself but all the ancillaries that take up so much time and so many resources – the media buildup, the pontifications of experts, and the constant taking of the temperature of public opinion, whether it is about tennis, football/soccer, snooker, horseracing, or politics. All this activity arises from something uniquely human – namely, the human interest in what other humans are thinking – and, in this case, backing their views with money.

In the AI economy, the same sort of thing will occur in many other, quite different, spheres. For several years now, over Easter the UK-based radio station, Classic FM, has run a survey of listeners' views on their favorite pieces of classical music, with the top three hundred pieces played from number 300 downward over the Easter weekend, and "the winner" being released late on Easter Monday. This has been tremendously successful, involving much chatter, analysis, and subsequent reporting. Note that none of this has involved an expert's analysis of the music or its worth, still less the creation of any new music. The whole exercise is simply about the testing of public opinion.

We can expect a proliferation of surveys of human behaviors and views on everything from sex to religion and drugs. Admittedly, most of the detailed work and the compilation of the results will be done by AI, but the design of the questions, their interpretation, and their presentation will be done by humans. Reality TV shows, agony aunt programs, accompanied by their non-celeb imitators in self-help groups, concentrating on the vicissitudes of human relationships, will boom.

Human relationships

In the Robot Age, now that people have more money and more time, they will surely be especially interested in enhancing their relationships with others, in several different respects. With the fight to live, the struggle just to get by, largely a thing of the past, people will surely put more emphasis upon becoming more beautiful, both to attract others and for their own satisfaction and sense of self-worth. And they will surely value beauty in others who interact with them, more so than ever. *Human* beauty. In this respect above all, humans will surely have a distinct advantage over robots, not to mention their disembodied cousins, AI.

This also extends to sexual relationships. Admittedly, attempts are being made to develop enhanced sex dolls that, thanks to AI, can interact better with their owners, both physically and in regard to "conversation." You could readily imagine a dystopian world in which human beings give up on the idea of a sexual relationship with other humans and rely instead on the artificial alternative. I suppose this is possible, but I suspect that sexual relationships with robots will remain a minority sport. Human beings have a need for intimacy – but with other humans.

People also have a deep need for friendship and companionship. Given that relationships of all sorts will be of continuing and perhaps even enhanced importance in the Robot Age, there will surely be a proliferation of jobs dealing with relationships – how to start one, how to keep one going, and even how to change or stop one. Relationship tutoring will surely be big business. And who would want to be tutored about how they should conduct their relationships with humans by a robot? Unless, of course, you are a robot. Well, are you? Even then, I suspect you will want, or be programed to want, to be tutored by a human.

In the Robot Age could there be an enhanced market for companionship? In the eighteenth and nineteenth centuries rich, single, older ladies would often pay for another, often younger, and certainly relatively impecunious, lady to be their companion. Even in today's world, I have known some examples of single, older men, employing younger live-in companions and helpers (and not for sexual reasons). They have wanted a combination of domestic help, security and conversation. Again, of course, you could employ a robot to talk to you and pretend to be human but that would not be "the real thing."

A vision of future employment

It is time to take stock. However, much we try to imagine what new jobs may emerge, one thing we can be certain of is that we will surely miss many of the new jobs that will spring up. How could it be otherwise? We are about to enter a whole new world and our imagination can only get us so far.

This is the key reason why assessments of the impact of robots and AI on total employment tend to be unjustifiably pessimistic. It is an easy, although laborious, task to trawl through lists of job specifications, assessing which types of job, and how many people, are "at risk" of redundancy. There have been umpteen studies of this sort. Doubtless there can be new studies, better or worse than their predecessors, and more or less apocalyptic. But, as regards the impact on the macroeconomy, these exercises are largely beside the point. The really important questions concern what modifications to existing jobs will effectively transform them and what types of *new* job will emerge, and in what quantities.

Not for the first time in economics, it is on the aspects of the issue that are most important that we know least. When thinking about our policies for the uncertain future, we must beware of designing them to deal with those things that seem to be quantifiable, while ignoring or underplaying those that don't.

Without saying that we can lay claim to much detailed knowledge, the above discussion ought to have put paid to the idea that may have plagued you that the new world must necessarily involve mass unemployment. In many sectors, including particularly transport, the scope for robots and AI to replace humans has been grossly exaggerated. In others – for example retail – there may well be significant job losses. And there will be many activities where the performance of routine mental tasks by humans are now replaced by AI.

But in many other sectors, including healthcare and the leisure sector, the employment of humans can expand dramatically. In some sectors, such as medicine and the law, robots and AI can enhance what professionals do. It is far from inevitable that this will lead to job losses in those sectors. Rather, as professionals' productivity improves, it is likely that their output will rise also.

Meanwhile, new jobs that barely exist today, will spring up and multiply. Accordingly, I see no reason why the robot- and AI-infused economy of the future cannot be accompanied by full employment.

Mind you, this is not necessarily a vision of the future that you will be bound to like. Quite apart from anything else, the destruction of many old jobs and the emergence of new ones will surely have implications for the makeup of society which could have decided political consequences – a subject that I take up in Part III.

But before we get to that, there is another potential concern that has not yet been laid to rest. Just because there will be enough jobs "to go round," this does not necessarily imply anything about what these jobs will pay. Will the new robot- and AI-dominated future necessarily imply a large number of people on low pay and therefore a substantial increase in inequality?

6
Winners and losers

> "In the coming century or so, humanity will divide into two classes of people – the gods and the useless."
>
> *Yuval Harari*[1]

> "Some people get rich studying artificial intelligence. Me, I make money studying natural stupidity."
>
> *Carl Icahn, billionaire*[2]

So far, we have discussed the impact of robots and AI largely as though the effects fall equally on all people and parts of the world alike. That is clearly not true, and it is time to set the record straight. I will now discuss the effects on the distribution of income across different individuals, groups, regions, and countries, starting with individuals.

One individual compared to others

To read many accounts of the impending shock to jobs and incomes about to hit us from the robot and AI revolution, you would think that this is something new – certainly something to be contrasted with the continued economic progress registered until recently.

In fact, although the aggregate statistics suggest that sometimes economies have been able to cope with and adapt to extraordinary technological change without serious ill-effects, hidden beneath the aggregates are painful human tragedies. Just because there are enough new jobs to replace the old ones that have been destroyed, this does not mean that particular individuals, or groups, or even regions and countries, have been able to switch easily to the new activities that are in demand.

Nor was this just a feature of the grimy, smokestack phase of the Industrial Revolution, back in the nineteenth century. Across much of Europe and North America the onset of de-industrialization and globalization in the 1980s and '90s devastated whole communities and regions. You can still see some of the effects today.

In 1980s Britain the economy was transformed by what is now called "The Thatcher Revolution." This involved the creation of many new jobs, mainly in the services sector. But it also involved the destruction of millions of "old" jobs, particularly in manufacturing and, most painfully, in coal mining. Many of the individuals made unemployed never really recovered. It was next to impossible to take a 50-year-old miner, who was almost without exception male, and transform him into a customer-facing call center operative. The people filling this role were, almost without exception, female.

And this problem didn't just affect isolated individuals. Because the economic activities hit by technological and policy change were geographically concentrated, whole communities and even regions suffered. Indeed, many communities and regions have still not fully recovered from the loss of their primary economic activity and source of employment.

Something similar is happening in America today. Among all American men, the rate of participation in the workforce (meaning those who are either in work or actively looking for work) has fallen from 76 percent in 1990 to about 69 percent in 2018. Moreover, since the late 1990s there has been a worrying rise in mortality among white, middle-aged, Americans, associated with alcohol, drug abuse and an increase in suicide.[3] It is likely that these two things are connected. Just as in the UK, the decline of traditional industries, accompanied by high levels of male unemployment and a proliferation of psychological and societal disorders, is geographically concentrated.

Even if I am right to argue that the coming AI revolution is going to be enriching for the human race, both materially and in other ways, the benefits will not be evenly spread. Indeed, it is likely that some people will be worse off, not just relatively but absolutely, just as happened in the early decades of the Industrial Revolution and has been happening more recently in most industrial countries as traditional sources of employment have fallen into steep decline. Indeed, in the AI economy this could be the fate of millions of people in America and other developed countries – even if there are "enough jobs to go round."

Before we come to discuss this possibility, we need to review the current situation because there is widespread anxiety about the distribution of income and what might happen to it, even before any effects from the AI revolution are felt. Moreover, the nature of the forces already making for increased inequality may interact with the AI revolution, as well as affecting what measures can and should be taken to address the issue (which I address in Chapter 9).

The facts on increased inequality

The place to start is with the facts. Or rather with what we think are the facts. For, like so many other critical issues in economics, this one is bedeviled by controversy.

Certainly, the conventional wisdom is that over recent years the distribution of income and/or wealth has been becoming more unequal. According to the Dutch historian Rutger Bregman, in the USA, the gap between rich and poor is "already wider than it was in ancient Rome – an economy founded on slave labor."[4] Whether or not this is an exaggeration, the bare statistics do show a marked increase in inequality.

In the USA, between 1962 and 1979, the annual average growth in real disposable income for people in the bottom quintile of the income distribution was almost 5.5 percent, compared to less than 2 percent for people in the top quintile. But between 1980 and 2014, for the bottom quintile, the average increase was next to zero while for the top quintile it was 2.8 percent. The top income quintile took home 44 percent of total post-tax income in 1980, with the top 1 percent alone taking 8.5 percent. By 2014 these shares had risen to 53 percent and 16 percent, respectively.

Not all countries have had the same experience. The UK, for example, has been quite different.[5] But let us take the US case. Why has inequality increased? Again, this is an area where economists have not stopped arguing (and they probably never will). But pretty much everyone agrees that there are two major factors at work: globalization and technological change. It is the relative importance of these two factors that causes the academic sparks to fly.

Globalization and inequality

The argument about globalization is straightforward. The opening up of China and other emerging markets to the world economy effectively added a couple of billion people to the world's labor force. The extra competition in the labor market was not spread equally across all categories. It was concentrated at the lower end where people had few skills. Accordingly, globalization tended to have a depressing effect on pay at the lower end in the developed economies.

Meanwhile, many people higher up the income distribution benefited greatly. Not only did they not suffer much direct competition from lower-paid people in the emerging markets, but they were now able to buy many goods and services more cheaply than before. In addition to the umpteen millions of people in the emerging markets who have been made much better off, these better off "Westerners" are the winners from globalization.

Over and above this, because globalization effectively added a couple of billion extra workers to the developed world's labor force but hardly any extra capital, it increased the returns to capital (profits) at the expense of the returns to labor (wages and salaries). Since share ownership is concentrated at the upper end of the income distribution, (despite widespread indirect ownership of shares through pension funds), this also increased inequality – at least in the developed countries. (Mind you, it is worth stressing that, on a worldwide scale, globalization has massively reduced inequality as it has enormously added to the income of millions of poor people in China and beyond.)

The technological twist

The technological explanation is also straightforward – but with an interesting, and important, twist. The straightforward bit is the continuing economization on the demand for labor, most recently thanks mainly to computers and associated developments.

The twist is that the communications revolution has caused the proliferation of so-called "winner-takes-all" markets. In traditional markets remuneration tends to be tied to absolute performance, but in "winner-takes-all" markets it is all about relative performance. One of the key factors that limited the extent of winner-takes-all markets has been distance. This has enabled a second- or even third-rate, or worse, provider of a service to stay in business and even prosper. But now, provided that the service in question can be digitized, then distance is no longer a barrier. This has effectively united the market globally and given access to the very best providers of a service in the world.

What's more, digital goods enjoy enormous economies of scale. This enables the market leader to undercut any competitor and still make a good profit. Once fixed costs are met, each extra unit produced costs hardly anything to deliver. The upshot is a tendency toward monopoly, with all the usual results. It is claimed that Amazon has almost 75 percent of the e-book market, Facebook is the medium for 77 percent of social media use, and Google has almost 90 percent of the market for search advertising.[6]

The new digital world produces winners on a scale never before imagined. The success of J. K. Rowling's Harry Potter books and films is one example. Another is the Korean pop song *Gangnam Style* and its accompanying dance. If you who are reading this have not seen the YouTube video of the song (never mind trying to do the dance), you are one of a few billion people who haven't. But it has been watched 2.4 billion times and it is still going. This size of audience is unprecedented.[7]

Great fortunes amassed in this way then lead on to second-line fortunes earned by people who supply the superrich. After all, if J. K. Rowling were ever to find herself in a legal dispute, she would surely employ the best lawyers going. She would surely not be interested in paying under the odds to secure the services of number two or three, let alone those far down the legal pecking order.

This is nothing new. The great economist Alfred Marshall (who taught Keynes) wrote: "A rich client whose reputation or fortune or both are at stake will scarcely count any price too high to secure the services of the best man he can get."

Erik Brynjolfsson and Andrew McAfee cite the example of the athlete O. J. Simpson who paid the lawyer Alan Dershowitz millions to defend him in court. Admittedly, Dershowitz's services were not digitized and sold to millions of people the way that Simpson's were. Yet, they say, "Dershowitz is a superstar by proxy: he benefits from the ability of his superstar clients whose labor has been more directly leveraged by digitization and networks."[8]

This talk about superstar athletes and authors may make it seem as though the winner-takes-all structure applies only in a very few isolated markets. But this impression is wrong. It applies across large parts of the economy. Why listen to the nth-rated orchestra when you can listen to the best? Why be taught by your local university's pedestrian-at-best professor when you can "be taught" remotely by the best professors that Oxford or Harvard have to offer? Why employ a tenth-rate accountant, investment banker, surgeon, or whatever, when you can employ the best? In short, why settle for anything less than the best that money can buy?

There is an answer to this question: it all depends. Where services are digitizable and there is no limit to the number of customers, then there is indeed no reason at all. In extreme cases, this will mean that any but the best providers of the service go out of business and there may be no price low enough to keep them in the game. As Brynjolfsson and McAfee put it: "even if we offered to sing "Satisfaction" for free, people would still prefer to pay for the version sung by Mick Jagger."[9]

This result does not quite hold where there are limits to the numbers of customers that a given provider can cater for. But still the result of extending the market is to drive up the market value of the top practitioners at the expense of the rest. Let's take surgery as an example. Robotics and AI may enable surgeons to operate on more patients, but the chief benefits come from increasing the safety and reliability of what they do and allowing them to operate remotely, perhaps even thousands of miles away from their patients. The result is that the demand for the top surgeons increases. Effectively they now face a global market. Less able surgeons will still be employed but at lower pay – unless and until robots and AI raise the quality and reliability of their work to that of the best.

The new Marx?

So, globalization and digitization have been two powerful forces operating over the last two decades to drive an increase in inequality. And now into this world of burgeoning inequality there steps a Frenchman bearing (intellectual) gifts. In 2014 Thomas Piketty published a book explaining the trend toward increased inequality in a different and powerful way and forecasting that it would intensify. Piketty's *Capital in the Twenty-First Century* became an international publishing sensation. It has proved to be the launching pad for thousands of books, academic papers, and PhD theses.[10]

Piketty's thesis is that the distribution of wealth and income is set to become ever more unequal because, quite simply, the return on capital exceeds the economic growth rate. This means that wealth grows faster than national income. Since wealth is highly concentrated, this is bound to lead to a steadily more unequal society.

Piketty claims that this has always been true in Western Europe for as far back as his data stretches. In the eighteenth and nineteenth centuries, he says, income was grotesquely concentrated. But the dramatic events of the early and mid-twentieth century – the First and Second World Wars, the Great Depression, the advent of the welfare state and progressive taxation – acted as correctives but also blinded us to the underlying reality. Now, Piketty says, the distribution of wealth has moved back to where it was in the late nineteenth century and, unless something is done to check this process, it is set inexorably to become more and more unequal.

Interestingly and, from the point of view of this book (i.e., mine, not his), remarkably, the word "robot" appears on only one of Piketty's nearly 700 pages. In other words, if you believe that robots and AI will dramatically increase inequality, *and* you accept Piketty's thesis, then we are heading for a very unequal world indeed.

Critiques of Piketty

As you might expect with a book whose impact has been as great as his, Piketty's thesis has come in for a welter of criticism. This is not the place to enter into a detailed account of these criticisms, but we do need to get the flavor of the key ones, and to reach a conclusion about the main issues, before going on to look at the impact of robots and AI.[11]

To begin at the beginning, while the sheer scale of the accumulation and analysis of data by Piketty has elicited widespread admiration, there are serious doubts about the quality and accuracy of this data. Indeed, these doubts are so serious as to call into question the whole edifice that Piketty has built on these foundations including, most importantly, the case for a top income tax rate of 80 percent and a global wealth tax of 20 percent or more.

There are three strands to this criticism. First, Professor Martin Feldstein has pointed out that Piketty relies on income tax returns for evidence about incomes.[12] Yet in the USA changes in tax rules since 1980 have given high earners less incentive to invest in low-yielding tax-exempt investments such as municipal bonds, and less incentive to put their business and professional income through corporations, which were taxed at a lower rate. Accordingly, although the tax returns may suggest increased inequality, in reality there may have been no underlying increase in inequality at all.

Second, Professor Herbert Grubel has pointed out that Piketty's data (such as it is) is focused on increased inequality between individuals at different points in time.[13] Yet, he explains, the same individual may make considerable movements between income groups over their lifetime. He cites Canadian data that show that out of 100 workers who were in the lowest income quintile in 1990, over the following 19 years 87 had moved to higher income quintiles. And 21 of these 100 individuals even reached the very top income quintile.

Third, Chris Giles, the Economics Editor of the *Financial Times*, has shown that there are serious problems with Piketty's data at a detailed level, including discrepancies between the data in original sources and Piketty's reproduction of them, as well as the crude insertion of "assumed" data where there are gaps in the original data. Giles says that: "The conclusions of *Capital in the Twenty-First Century* do not appear to be backed by the book's own sources."[14]

Theory and evidence

Given how devastating these criticisms of Piketty's data are, you might well think that it is superfluous to comment on the theoretical defects of Piketty's thesis. Nevertheless, we should briefly consider the theory. After all, there is a well-known quip in economics along the lines of "OK, so the idea doesn't work in practice, but does it work in theory?" Indeed, because disputes about data may continue ad nauseam, often the central idea to which they relate can take on a life of its own, even if it has no basis in empirical fact.

One of the leading theoretical critiques of Piketty's thesis is that as capital becomes ever larger in relation to other factors of production, and in relation to national income, it should become more and more difficult to earn a good return on it. (Economists refer to this as the law of diminishing marginal returns.) This is at odds with Piketty's idea of the return on capital continuing to exceed the economic growth rate pretty much forever.

The Nobel laureate Joseph Stiglitz has a fascinating and powerful explanation of how what should be diminishing returns on capital squares with Piketty's data. He says that we should not equate wealth with capital, as Piketty does. In particular, he says that in the modern world a good deal of the increase in wealth is down to the increased value of land (and other rent-producing assets). But a higher value attributed to land (because of increased demand) does not increase the amount of land, or other productive assets, available. Indeed, Stiglitz argues that over recent decades, as measured wealth has been rising, the amount of productive capital deployed in the economy may have been falling.[15]

Another major criticism is that the simplicity of Piketty's analysis rests on the assumption that all of the return on capital is reinvested and none of the income earned by labor is saved (i.e., used to build up capital). Yet in a modern economy, by contrast, much of the return on capital is spent on consumption while a good deal of the returns to labor are saved. This is particularly true in a world of pension funds and widespread property ownership.[16]

A further, telling, criticism is not theoretical but empirical. When you look at the rich lists, particularly in the USA, but also elsewhere, it is striking that so many of the superrich have acquired their wealth through work, leading to fantastic increases in the value of the companies that they founded and/or built up. This applies to Jeff Bezos, Warren Buffett, Bill Gates, Mark Zuckerberg, and many more.

And the key players are constantly changing. Consider the *Forbes* list. Of those listed as the wealthiest Americans in 1982, less than a tenth were still on the list in 2012. Moreover, the share of the *Forbes* 400 who came to their wealth through inheritance seems to be in sharp decline.

One of the interesting things to come out of Piketty's study is that the recent increase in wage inequality is very concentrated at the top. In the USA the gap between the incomes of skilled and unskilled workers appeared to stop increasing in about 2000. By contrast, the incomes at the very top have taken off. Yet it is difficult to explain how the ultrarich have got richer as a result of technology or employing foreign cheap labor. Piketty estimates that 60–70 percent of the top 0.1 percent are chief executives and other senior corporate managers.[17]

These features stem from factors quite independent of anything to do with AI – or with Piketty's theory. They arise from the failure of institutional shareholders to constrain the pay of senior executives and the financialization of the economy in the Anglo-Saxon countries. Already there have been moves to reverse both these factors. Since the Great Financial Crisis, banks have become more tightly regulated. More generally, institutional shareholders have started to become more active in constraining executive pay. But in both regards, there is room for much more progress. (More about this in Chapter 9.)

What is the upshot? Many on the Left hold up Piketty as a hero and adhere strongly to his theory that suggests that inequality is bound to increase inexorably. They see this making the case for radical public action, even before any effects from robots and AI. They can view the redistributive impact of the new industrial revolution as just something else to add on to, and perhaps speed up, Piketty's process of wealth concentration, or simply something to enable them to jump ship to a new leading cause of increased inequality if they feel that Piketty's thesis has some serious holes in it.

And, as we discussed above, it does have some serious holes in it. Scholars seems united in admiration for the work Piketty has done to assemble data on a range of countries over such a long period.

But few find his theory convincing. And nor do I. We do not need Piketty's simple but potentially powerful theory to explain what has happened to inequality. What's more, many of the facts on the ground seriously challenge his thesis. Accordingly, we can now leave Piketty to the economic historians to use as a database – and the foundation for umpteen keenly fought academic disputes.

Jobs and pay

So now we can turn our attention to the impact on the distribution of income of the AI revolution. Noneconomists often talk about employment in absolute terms. They refer to some jobs *disappearing* and others taking their place. Indeed, I used this simple approach in the discussion in the previous chapter.

But the natural method of economics assumes that the system is flexible. And the oil for this flexibility is price. Accordingly, it is better for us to think about the labor market in an AI- and robot-dominated world in relation to supply and demand and to see changes in these two fundamentals as being expressed in changes in both quantities and prices. In the case of employment, the price in question is wages and salaries. And, of course, what happens here feeds into the income distribution.

So who will the winners and losers be? Kevin Kelly, the editor of *Wired* magazine, has made a key contribution to discussion of this issue. He has said: "You'll be paid in the future based on how well you work with robots."[18] Google chief economist Hal Varian frequently says that people should seek to be an "indispensable complement" to something that's getting plentiful and cheap. Bill Gates has said that he decided to go into software when he saw that this was going to happen with computers.

In practice, as I argued in the previous chapter, this does not mean that you have to be an AI geek in order to succeed in the AI economy. Indeed, I argued there that so many of the jobs on offer – and good jobs at that – will center on the human factor. How well you interact with human beings will be at least as important as how well you interact with robots, and perhaps more so. The losers in the AI economy will be those who are doing essentially robotic jobs and/or are themselves irredeemably robotic when dealing with other people. I am sure that readers will know umpteen examples of these human types from their own experience.

It is tempting to assume that the losers will be people at the bottom of the income and educational heap who will be outcompeted by robots and AI. If they manage to find jobs at all, they will surely be at very low incomes. Indeed, their worth in the market may be so low that they fall below a decent level. Accordingly, millions of such people may choose, or be forced, to rely on state benefits. Meanwhile, many, if not most, of those higher up the income and skill scale will be protected from robots and AI, in the same way that people on high land are protected from rising flood waters.

Actually, I am not at all sure that this vision is correct. Earlier rounds of mechanization, and indeed computerization, and even the industrial use of robots, were indeed centered on the replacement of unskilled or lowly skilled manual labor in factories. Professional and even clerical jobs at first saw little impact. But subsequently they did experience a major impact. Indeed, the jobs of secretary and filing clerk effectively disappeared. In most Western countries this process is now almost complete.

As I argued in the last chapter, many of the jobs now at risk from AI are moderately skilled nonmanual positions, which are often reasonably well paid. By contrast, the demand for basic manual labor is likely to hold up well, and consequently people doing these jobs should still be (moderately) well remunerated. Moreover, the AI revolution can, in some ways, improve their position as the demand for their services increases with increased income and wealth.

There is a widespread presumption that, whereas the majority of the jobs lost to technology are relatively highly skilled and well paid, new jobs tend to be low-skilled and low-paid positions, predominantly in the service sector. We have a mental picture of the new jobs on offer as low-paid work in fast-food restaurants or as Deliveroo riders. But in fact, the average skill content of UK jobs has continued to increase. Over the last 10 years the share of the three highest-paying occupational categories (managers and senior officials, professional occupations, and associate professional and technological) in total UK employment has risen from 42 percent to 45 percent.

Moreover, in even the most supposedly unskilled of activities, workers in fact need mastery of several basic skills which their equivalents in previous generations would not have had and wouldn't have needed: the ability to use and communicate by telephone and computer, and the ability to drive a car. It is true that there is a significant number of members of today's society who do not have these skills, but the majority of people do have them. What's more, these skills will not usually have been learned in any formal way but will have been acquired as a normal part of modern living. The ability to drive, or at least to pass the legally required driving test (which is not quite the same thing), is an exception. The result is that workers have more skills – that is, if you like, they possess more human capital – than may appear at first sight.

How AI can reduce inequality

Some AI advances produce surprisingly favorable results for people on lower incomes. The Uber taxi service has transformed travel in many cities. In London it has undermined the famous black taxi trade. The drivers of these iconic cabs famously spend great time and effort learning the streets of London – "the knowledge" – by heart. But the worth of this knowledge has been radically reduced by the availability of satellite maps that can be received on a smartphone. Accordingly, black taxis are regularly undercut by drivers who know next to nothing about the city that they are driving around.

From a purely economic point of view, this increase in supply has the result of reducing the equilibrium price. Yet if you try to suggest to a London cabbie that black cabs are too expensive and the black taxi trade ought to lower their fares (which are set by a regulator), you will be lucky to escape with your life. Your cabbie will take it that you are implying that he "deserves" to earn less than he customarily has. He will give you a very rum response, which cannot be committed to print.

Yet because the price of taxi rides provided by Uber is much lower, they have greatly expanded the market for taxi rides. And because the technology enables Uber drivers to identify and collect passengers quickly and efficiently, they spend less time than traditional taxi drivers sitting idle or cruising the streets hoping to pick up "a fare." The result, according to some, is that Uber drivers earn more than traditional taxi drivers. In the USA, it is claimed, they earn about $19 per hour compared to about $13 for traditional taxis.[19]

Admittedly, if driverless cars eventually proliferate, this increase in incomes for at least some commercial drivers may prove to be merely transitory. Yet, as I argued in the last chapter, the impact of driverless vehicles on the demand for human drivers is likely to be much smaller than the enthusiasts allege.

The Uber example provides an interesting window on the general effects of technology on the income distribution. Uber's advent transfers income from old-fashioned taxi drivers to the lower-skilled drivers of Uber cars. Does this increase inequality? Hardly. Traditional taxi drivers have tended to be relatively well paid compared to other working-class people. Indeed, you could say that this transfer is making the distribution of income more equal.

Then there are the benefits to consumers. Are these skewed toward the better off? Not necessarily. In fact, probably the reverse. And how do you value the increase in travel (and greater convenience and safety) now made possible by the new technology? Surely this benefit will be felt most keenly by people toward the lower end of the income scale, whose demand for travel is much more price sensitive and who, in the past, have been unable to afford to travel by taxi.

Admittedly, the Uber revolution does potentially bring an element that could make the income distribution more unequal, namely the potential earning of huge profits for the owners of Uber, who will not be evenly spread across the income distribution. They will surely be among the better-off members of society. But, so far at least, this point remains of only theoretical interest, since Uber currently makes a thumping loss.

The benefits of cheaper services

The Uber phenomenon is not an isolated example. Most of today's disruptive technologies are eroding the cosseted position of professionals, skilled workers, and established businesses, thereby benefiting the whole population.

This is especially true of the fintech industry, which is eating away at formerly huge margins and spreads in many areas of traditional banking, including foreign exchange, deposit-taking, and lending. The effect is to spread the benefits of the lower costs of providing these services across the whole income distribution. Besides the owners of shares in banks, those losing out will be the various bank workers who have found employment, usually extremely well rewarded, in the banks' traditional intermediary roles.

Something similar may be true of basic legal services. They will become much cheaper thanks to the employment of AI. The result will be increased use of such services. The beneficial effect will fall disproportionately among the less well-off who, on the whole, currently find legal services prohibitively expensive.

Accordingly, it is by no means obvious that the AI revolution is bound to increase income inequality. Indeed, it is possible that, at least across some parts of the income distribution, the effect of the AI revolution will be to *reduce* it. After all, the thrust of preceding chapters is that many manual jobs will not readily succumb to automation. Meanwhile, many skilled but essentially routine white-collar jobs will. Prime examples of the latter include large numbers of mid-level lawyers and accountants. Such people have typically earned much more than the average manual worker.

Mind you, this does not settle the matter. In the USA there has recently been an increase in demand for the skills of those at the very top of the distribution and also an increase in demand for services of the large number of relatively unskilled people at the bottom, at the expense of those in the middle with moderate skills of a mechanical variety – of the sort that can readily be replaced by AI. But this does not appear to have done much to boost wages at the bottom because there is a continual flow of people from the middle skill and income range into lower skill and income occupations.

By contrast, there is not much of a flow of displaced middle-class people upward because moving into this group generally requires a much higher level of education, including professional qualifications, that can't just be acquired overnight. It may take several years of training before someone can move into this higher group – even if they are capable.

In the literature, the resulting phenomenon of more jobs at both the top and the bottom of the income distribution phenomenon is known as "job polarization." This seems to augur badly for the chance of increases in incomes for people at the bottom of the income distribution as their incomes continue to be depressed by this constant flow of people displaced from above. But David Autor, a professor of economics at MIT, thinks otherwise. He suspects that most of the erosion of middle-income jobs because of computers and AI is now over. What we are left with now is many occupations where AI is complementary to human skills, including radiology, nurse technicians, and other medical support jobs.[20]

Autor notes that journalists, commentators, and even AI experts typically overstate the extent to which machines substitute for, and understate the extent to which, they complement human labor, thereby missing the scope for increased labor productivity and higher real wages. This is as true for AI as it was for computers and earlier waves of mechanization.

But this generally means that workers have to upskill to some degree in order to be able to interact productively with AI. Again, much the same is true of computers. There can now be hardly any office jobs, indeed any jobs, that can be performed today by someone who cannot operate a computer. Construction workers who can wield a shovel but not operate a mechanical digger may still be in some demand, but they earn much less than the worker who can both wield a shovel and operate a mechanical digger.

Technology and monopoly

I argued above that in the digital world and, by extension, in the new AI-infused world, there may be an inherent tendency toward monopoly which would tend to lead to increased profits thereby, other things equal, tending to increase inequality as the ownership of the companies enjoying these monopolies will be skewed toward the rich.

But this argument cannot be accepted uncritically. Although the digital world does indeed seem to be characterized by a high degree of oligopoly or monopoly, its abiding feature is disruptive change.

After all, iconic companies like Facebook, Google, and Amazon did not exist 25 years ago. Their emergence from nowhere eroded or destroyed the profit position of many established businesses and brought great benefits to consumers.

Accordingly, it would be most unwise to assume that their own market position is set in stone for all time. Because of economies of scale and first mover advantage, they may not be very vulnerable to competitors trying to do the same as them – although they cannot be sure of this. Their real vulnerability is that some new technology may come along and undermine them, just as happened to those who preceded them.

Furthermore, some modern technologies, such as blockchain and 3D printing, facilitate small-scale production.

Moreover, another surprising possible effect on the income distribution springs up from the discussion of work versus leisure in Chapter 4. It transpires that, in a turnaround from most of human history, in today's society, those at the top of the income distribution currently tend to work longer hours that those at the bottom. There is some evidence that this does not accord with people's preferences – in either group.

Now, suppose this is right and suppose that in the economy of the future people at the top end work less, while people at the bottom end work more. This would naturally tend to reduce the disparity between incomes (measured in money, not taking account of leisure time) between top and bottom.

Admittedly, though, the robot and AI revolution may well favor capital over labor. And since capital is mainly owned by the better off, in the absence of intervention, this is likely to lead to a skewing of the benefits of robots and AI towards those at the top of the income pile.

This aspect could have a decided intergenerational aspect. The owners of capital will be predominantly old, while the sellers of labor will be predominantly young. Accordingly, it is easy to see the changes unleashed by the AI revolution as benefiting the old at the expense of the young.

And there could potentially be a major impact on social mobility. If there is widespread impoverishment, whether because jobs are scarce or because they pay very little, then it will become extremely difficult to accumulate capital out of income.

Overall assessment

It is understandable that many people fear a future distribution of income in a robot- and AI-dominated world where the mass of people are scrambling to get low-paid jobs. Meanwhile, there is a swing in favor of the owners of capital and those few humans who possess the skills in demand in the new world. In these conditions many people would face a choice between low-paid, and probably irregular and precarious, employment and long-term unemployment.

This effect would come on top of powerful forces in the modern economy that have already been pushing in this direction. Certainly, many people, including many of the great figures of the digital and Artificial Intelligence worlds, fear such a future.

If something like this does transpire, then it may well require significant intervention by the state to prevent and/or correct it. This runs into questions about the ideal, or at least acceptable, distribution of income in society, as well as the costs and unintended consequences of measures that might be taken to correct increased inequality. I will discuss these issues in Chapter 9.

Yet it is far from being a foregone conclusion that the AI economy will involve much increased inequality. Indeed, many of the developments in the AI revolution will be pushing in the direction of reduced inequality as they undermine middle class incomes by providing services at lower prices, thereby benefitting people lower down the income scale. Of course, many jobs will be lost but new ones will take their place. Nor is it inevitable that these new jobs will be low-paying. The ability of the great mass of people to obtain relatively well-paid jobs will depend upon a series of factors:

- The technical capability of robots and AI. (At the moment across a broad swathe of activities, where they act independently of humans, their performance has been disappointing.)

- The cost of making, maintaining, and financing robots and AI.
- The scope for robots and AI to interact advantageously with humans (the more that robots and AI serve as complements to human labor rather than substitutes for it, the higher will be the wage than humans can earn.)
- How much humans want to buy things and, more especially, services, from other humans.
- How far robots and AI reduce the cost of services currently provided by the relatively well paid to the benefit of consumers who are less well paid.
- The degree to which human beings voluntarily decide to take more leisure. (The extent to which this happens will strengthen the bargaining position of human labor versus capital and AI.)
- The extent to which the preference for more leisure is skewed toward people at the top end of the income distribution.

Clearly, all these things are interactive, and it is extremely difficult to tell how things will pan out. But it is likely that the balance of the factors listed above will enable the mass of people to enjoy increasing incomes even as robots and AI start to proliferate. And it is perfectly plausible that there will not be a significant increase in income inequality.

If there is, it may well be a temporary phenomenon. The great economist Simon Kuznets argued that economic development would at first increase inequality but that subsequently this widening would be reversed. Moreover, this story fits in with the history of the Industrial Revolution. As I showed in Chapter 1, in the first few decades of the nineteenth century the real incomes of workers fell.

The upshot is that it is far too early to conclude that as robots and AI take hold there is bound to be a major move in the direction of increased inequality. We simply don't know. After all, if we had known at the end of the nineteenth century that employment in agriculture, tending horses, and domestic service was about to plummet, it would have been easy to conclude that the outlook for the less well-off was about to deteriorate markedly and to imagine the swollen ranks of the poor scrambling around for any scraps of work and income. The outturn, of course, was exactly the opposite. Why could something similar not happen again?

Winning and losing regions

Is the same true of regions and countries? You might readily assume that, since robots have no preference for where they are located and, at least until they take over the world (to be explored in the Epilogue), would not be taken notice of even if they did, while AI can surely be employed anywhere, the coming revolution undermines the advantages of geographical concentration and agglomeration.

This could readily be taken to mean that economic activity can be located anywhere. Accordingly, this might imply that regional disparities will be reduced as economic activity, and hence accommodation for people, might as well migrate to where land and services are cheaper. The implication would surely be a migration away from expensive cities toward cheaper towns, villages, and regions.

Yet I am wary of accepting these conclusions. It would not be the first time that "the Death of Distance" has been pronounced – prematurely. This was supposed to happen as a result of the computer and communications revolution as people were now able to work, play, and communicate at great distance from each other. To some extent this has happened, of course. But it has not led to the dispersal of economic activity. Indeed, the tendency toward agglomeration looks to have continued.

London is a prime example. You might easily have imagined that the communications revolution would have diminished London as a financial center and boosted provincial centers within the UK, such as Edinburgh, Glasgow, Dundee, Perth, Birmingham, Norwich, Bristol, Leeds, Manchester, and Exeter, all of which used to have significant amounts of financial activity, and all of which I used to visit, in my youth, as a financial market analyst. In fact, though, these places have seen locally headquartered financial businesses being merged or taken over, with activity moved to London. Only Edinburgh manages to survive as a significant financial center, albeit at a fraction of London's size.

Interestingly, the equivalent is not true *within* London. It used to be said that to be anyone and achieve anything in the financial sector you had to be located in the heart of the City of London, within walking distance of the Bank of England. But if this used to be true, it certainly isn't now. The financial services industry has split. Large numbers of institutions are still located in the old City, but many businesses have now migrated to Canary Wharf in the east, while many others have moved westwards, to the cluster around St James's and Mayfair, the home of hedge funds. There are also offshoots in Victoria and Marylebone.

What could explain these two contrasting trends? I think it must be because although much business can be done remotely, this has, if anything, served to increase the significance of face-to-face meetings. If you are based in hedge fund alley in St James's, you can readily arrange to meet at short notice someone who is based in the City, or even Canary Wharf. But you cannot do that with someone based in Birmingham, let alone Glasgow.

Similarly, firms operating in these different parts of London are effectively able to draw upon the same, extremely large, pool of skilled labor. By contrast, the labor forces of Glasgow, Birmingham and greater London are, to a large degree, segregated.

Similar forces making for agglomeration are also at work in the world of leisure. Football games, horse-racing venues, theater, opera houses, concert halls, top restaurants all need to be located within easy reach of their clientele. And this has a clear knock-on effect on the location of much economic activity – and not just all the things involved in the provision of those various entertainment services. If you are a leading hedge fund manager with considerable disposable income to spend on restaurants, shopping and/or the opera, would you rather live in Scunthorpe or London? (For the USA, read Peoria, Illinois, versus Manhattan.)

It is said that this factor has had a marked bearing on the continued presence of large numbers of hedge funds in London even after certain unfavorable tax changes made a relocation to Switzerland seem more alluring. Geneva certainly has its attractions. But on the range of shopping, music, and restaurants it cannot hold a candle to London. And then there is the Swiss provincial city of Zug. (Nothing against Zug, by the way. Some of my best friends have been there.)

Professor Ian Goldin of Oxford University has argued that the changes unleashed by the AI revolution will actually *increase* regional disparity. The most vulnerable towns and regions are geographically isolated from the dynamic cities experiencing strong growth. Moving to these cities is inhibited by high housing and commuting costs.

A similar conclusion is reached by the think-tank The Centre for Cities. It expects that by 2030, under the twin impact of automation and globalization, about 3.6 million existing jobs in British cities – that is about one in five – will be displaced. It thinks that the job losses will be heavier in the North and Midlands. Meanwhile, in the South, cities are already ahead in things like software development, which will enjoy increased demand, and are likely to suffer smaller losses proportionately.[21]

A report for the IPPR Commission on Economic Justice concluded that, within the UK, London has the highest proportion of jobs that are likely to be resilient in the face of automation.[22] Interestingly, this report also suggested that women and some ethnic groups were disproportionately employed in jobs that are susceptible to automation.

The author and AI visionary, Calum Chace, suggests that people will increasingly cluster according to income. The hothouse cities of San Francisco and New York will be no-go areas for most of the poorer 85 percent of people. He says that they will inure themselves to the glaring inequalities in society with the "opiates of free entertainment and social media."[23]

Personally, I am not so sure about any of this. Just as with the impact on the distribution of income, we need to be wary of jumping to conclusions when so much of the future is clouded in pure uncertainty. We must be prepared to be surprised.

Winning and losing countries

So far, I have implicitly assumed that the robot and AI revolution sweeps over all countries equally. Yet a moment's reflection should establish that this is most unlikely to happen. The new world into which we are gradually passing is so different from the old one and the challenges that it poses for individuals, companies, and governments are so intense, that we must assume that different countries will make different choices and be differently successful in implementing them.

Just as the Industrial Revolution revolutionized the international balance of power, this new revolution could do the same. The Industrial Revolution launched the UK into greatness because it industrialized first and for a while, until Germany and the USA overtook it, it enjoyed a considerable first mover advantage. Is there a country that is clearly placed to be a leader in the Robot Age? And are there countries that we can readily identify as likely losers?

It should be stressed that taking full advantage of AI comes in two forms – production and consumption. These will not necessarily go hand in hand. It is perfectly possible that only a few countries will take a leading role in the production of AI-enabled products and even services – probably, the USA, China, and perhaps in some fields the UK – and yet these products (and services) may be employed in many more countries.

This mirrors the development of computers. Very few countries are engaged in computer manufacturing, and the development of the appropriate software is heavily dominated by the USA. But, of course, computers are used everywhere in the world. Moreover, if a country decided to eschew the use of computers because it did not produce them it would consign itself to the economic scrap heap.

The same is true of AI. Just because your country does not produce AI – none of the algorithms, deep learning apps that are driving AI, nor the physical entities, such as robots – this does not mean that you cannot benefit by employing them. Indeed, if you don't, you risk falling into economic irrelevance.

That said, there is a marked difference of opinion among the technological cognoscenti about how innovation, including with regard to AI – and the gains from it – will be distributed globally. At one extreme, there is the "world is flat" view, advocated by the *New York Times* columnist Tom Friedman. This holds that, thanks to the cheapness and interconnectedness of modern technology, anyone with internet access can invent a service or product with potential global scale.

At the other extreme is the view that the world is far from flat. In fact, it has at least a few large mountains. This is the view of Professor Richard Florida from Toronto University, who emphasizes the fact that innovation takes place in a select number of metropolitan areas, usually centered around a large, successful company and/or a leading university.

Yet, although the world seems to be dominated by American tech firms, at the start-up and just past start-up stage, innovation has been globalized, with vibrant tech centers developing in China, India, and Europe. According to a report jointly written by Professor Florida, the share of Silicon Valley in Venture Capital funded deals is now not much more than 50 percent, compared to 95 percent in the mid-1990s.[24]

But the study also reveals that the top six cities attracted over half of all venture capital investment and the top 24 attracted more than three-quarters. Strikingly, three of the 10 top cities by invested capital are in China.

A 2018 report from the OECD concluded that in general jobs in Northern Europe and North America are at less risk than jobs in Southern and Eastern Europe. It concluded that in West Slovakia almost 40 percent of jobs are at risk. By contrast, in the area around Oslo in Norway, it says that only about 4 percent of jobs are at risk.

These differences are essentially to do with the current structure of employment. But, going forward, the key characteristics making for relative success between countries are likely to be different: the amount that is spent on AI development and research; the degree to which robots and AI are tightly regulated and/or taxed; and cultural factors governing how readily robots and AI are accepted by people at large.

Differential spending

There are already huge differences in the intensity of robot employment across countries. In 2016 the highest number of industrial robots per 10,000 employees in manufacturing was registered by South Korea with 631. Thereafter, the ranking of some key countries goes as follows: Singapore (488), Germany (309), Japan (303), USA (189), Italy (185), France (132), UK (71), China (68), Russia (3).[25]

And different countries are investing very different amounts in AI. According to Goldman Sachs, between the first quarter of 2012 and the second quarter of 2016, whereas the UK invested $850 million, China invested $2.6 billion, and the USA roughly $18.2 billion. Despite being so heavily outgunned by China and the USA, the UK was the world's third highest spender. It is noteworthy that China has committed itself to becoming a world leader in AI by 2030. By then it aims to have built up its "AI ecosystem" to $150 billion.[26]

AI is often thought of in terms of robots or drones, but its impact will often be greatest from a less visual source, namely the ability to analyze data. And China potentially has a massive advantage in this area – namely much more data. The signs are that China is going to be one of the largest employers of AI. In 2017 China filed for 530 camera and video surveillance patents, which was more than five times the number applied for in the USA. Chinese people can already use their faces to shop, pay, and enter buildings.

It is noteworthy that, whereas in the US AI research effort has largely been related to the internet, in Germany research has been concentrated on areas where AI can improve manufacturing processes and the production of "smart goods" that integrate AI, such as cars and domestic appliances. It is also noteworthy that, in keeping with both its industrial and intellectual history, the UK's AI research effort has flourished mainly without any attempt by government to determine, or even influence, its direction.

Yet in April 2018 the UK government announced that it was investing £300 million in AI research. This was meant to sound impressive but in fact this is peanuts compared to what is being spent internationally. In written evidence to the House of Lords Committee on Artificial Intelligence, Microsoft noted that "in a year when China and India each produced 300,000 computer science graduates, the UK produced just 7,000."

Admittedly, the UK has an outstanding record in research into AI but, as in so many fields, it has been less successful in turning this prowess into commercial success. Moreover, there has been a marked tendency for small and innovative British AI firms to be sold out to large, foreign (usually American) businesses. The sale of DeepMind to Google is an obvious example. Founded in September 2010, it was sold in 2014. And whether or not UK-founded firms are sold abroad, much of Britain's top AI talent decamps to other countries, mainly the USA.

There is a serious prospect of the UK trying to cover the waterfront in AI, as in so much else, and failing, as it hasn't a hope of matching the scale of China's or America's spending. Instead, there is much to be said for specializing, as Germany seems to be doing. A report from the same House of Lords Committee on Artificial Intelligence mentioned above seems to think that the UK could become a global leader in the ethical and regulatory aspects of AI – although quite how this translates into major economic success is unclear.[27]

Yet the area of the UK's AI specialization could be much broader. Given that AI will be complementary to many types of skilled human labor, it would not be surprising to see the spread of country specialization in the application of AI mirroring the cross-country distribution of such human skills. Hence the USA and the UK might specialize in the application of AI in the fields of finance, law, accounting, business services, and, perhaps, medicine. Germany, China, and Japan might conceivably lead in the application of AI to advanced manufacturing.

Meanwhile, on recent form, you wouldn't readily assume that many countries in Africa would manage to invest sufficient capital, or manage the radical organizational changes, required to make a success of the AI revolution. If they did, their prospective area of specialism might possibly be in the application of AI in basic manufacturing, mining, and agriculture.

Mind you, just because a country starts off with a "lead" and a specialism in a particular field, this does not necessarily mean that it is bound to retain that lead and successfully manage to develop a specialism in the application of AI to that field. Relative competencies and specialisms can and do change. And if a country does badly in incorporating AI into the activity in which it has developed a specialty, then it is liable to lose its specialism to other countries that manage to incorporate AI better.

Inclination to regulate

On past form, as well as on the basis of current amounts of money being spent on AI, you might readily expect the USA and China to be in the forefront of AI and related developments. Moreover, they are the countries which are most likely to allow the economic changes implied by the AI revolution to have their head, without moderating or suppressing them for social reasons. Ocado told the same House of Lords Committee that China's relative lack of regulation gave it an advantage, commenting that "less regulation around the application of technology is fueling faster experimentation and innovation, including when it comes to the use of data and AI."[28]

By contrast, you can readily imagine some countries taxing and regulating robots and AI more heavily than others. In general, those are likely to be the type of societies that are already prone to giving close protection to the interests of existing groups of producers. The EU readily springs to mind as a likely heavy taxer and regulator of robots.

It is impossible to be confident about where the UK would fall on this spectrum. Much depends upon how it seeks to position itself in the post-Brexit world. I guess it will probably be somewhere between the USA and the EU.

I discuss how countries may choose to regulate robots and AI and the consequences thereof, including the issue of the possible introduction of a robot tax, in the next chapter.

Cultural factors

Tax and regulation aren't all that matters here. Culture is also important. In general, Asian countries, especially China, Japan, and South Korea, seem to embrace robots much more readily than Western ones. In general, Asian people do not think of robots as threatening. Generations of Japanese children have grown up thinking of robots as helpful heroes. *Astro Boy* is a manga series that has sold 100 million copies. By contrast, in the West robots are seen as deeply threatening. People's inner prejudices about them are strongly influenced by the *Terminator* movies.

Perhaps this difference has deeper cultural roots. Masatoshi Ishikawa, a Professor of robotics at Tokyo University, thinks so. He suggests that the fundamental reason for the difference is religious. He contrasts monotheistic religions of the West that cannot credit a major nonorganic entity with intelligence with the spiritualist religions of the East. Adherents of the latter can find it easy to believe that robots have a "spirit."

Whether the roots of it are religious or not, Western literature is full of stories about human beings creating things that they then cannot control. Perhaps the most famous of these is Mary Shelley's *Frankenstein*. The result is that robots and AI are likely to be less readily accepted by the public in the USA and Europe than in Asia. This could potentially influence these societies' inclinations to heavily regulate and tax robots and AI.

Different national choices

If, as argued in this book, the spread of robots and the employment of AI promise to usher in a dramatic increase in productivity, then the relative ranking of countries in the GDP league could shift appreciably depending upon how readily robots and AI are embraced rather than taxed and heavily regulated.

Similar consequences would flow from different choices that countries make with regard to the trade-off between work and leisure. If one country decides to take the AI dividend in the form of increased leisure, whereas another decides to take it in the form of increased real output and income, then the economic weight of the former will fall relative to the latter.

In many ways, this might not matter, but there is one respect in which it certainly would matter, namely defense capabilities. Other things equal, if one society opts for more leisure and another opts for more output and income, then, over time, the second will increase in relative defense capabilities. I say defense, but I could perhaps more accurately say *offense*. The same argument applies if different countries adopt very different policies with regard to the taxation and regulation of robots and AI.

Here again, though, there could be a policy response. The worry that a country could find itself way behind in its capacity for spending on defense, thereby influencing its strategic vulnerability, may operate on its leaders to persuade them to slow down or counteract any tendency for their society to choose more leisure over more output, and perhaps also to relax any inclination to excessively tax and regulate robots and AI.

A spanner in the development process?

One implication of the AI revolution may be to make it more difficult for developing countries to progress to the front rank of economic development. Over recent decades a string of countries, particularly in East Asia, have made this transition based on low labor costs, enabling them to expand their manufacturing sectors through heavy reliance on exports. This is the route to advance taken by Japan, Korea, Taiwan, China, Singapore, and Hong Kong. These countries were able to register impressive increases in exports, thereby enabling them to benefit from economies of scale.

At first, this was restricted to manufacturing, but with the communications revolution it spread to parts of the service sector. Call center work and basic accounting and legal work have increasingly been transferred to cheaper centers abroad, particularly in India.

But in the new world of the robot and AI, labor costs will be of diminished relevance. If you can get goods manufactured cheaply with little human input, why locate manufacturing in Asia when you can manufacture locally and thus avoid transport costs and delays? You could readily imagine a reversal of the trend of the last 30 years whereby much of Western manufacturing migrated to the East. (If anything like this were to occur, one of the largest categories of loser would be major international shipping groups.)

Similarly, why employ cheap but still non-zero cost human labor in India to do administrative tasks, with all the problems this may produce, when you can instead employ an app which costs next to nothing? Admittedly, sometimes the switching of digital services between locations can have some surprising features. Recently, there has been a move to locate machines in *colder* climates – where the costs of keeping the servers cool are lower.[29] Apparently, Iceland is a favorite location.

Of course, the driver of international trade is differences in *relative* costs, as outlined by David Ricardo in his theory of comparative advantage 200 years ago. So, it will always be advantageous to trade rather than not to trade. Nevertheless, this principle does not establish the amount of trade that will be profitable (and desirable). And it may well be that in the new economic conditions, thinking in narrow terms and at the overall global level, a reduction in the amount of international trade would be beneficial.

But would this destroy the path to prosperity for many developing countries? It would probably do nothing to inhibit the development of East Asian economies. For them the bird has already flown. Many of them have already attained living standards at, or at least close to, those enjoyed in Western developed countries. Admittedly, China and India are not there yet, but they have massive internal markets which should allow them to bypass reliance on exports as the route to success. Moreover, in both countries, but especially India, labor is still cheap, and it will still be profitable to employ human labor rather than robots or AI in a wide spread of economic activities.

The serious losers could be those countries that have not yet managed to make much progress up the development ladder. The economist Dani Rodrik has warned of "premature deindustrialization," as countries still low on the development ladder are prevented from industrializing through exports and are driven to become service economies.

Many African countries readily spring to mind. It has been common to argue that some of these could be about to follow the same development path trodden by so many East Asian countries. After all, as East Asian countries have developed, so their labor costs have soared. This has made it possible to imagine that just as Japan transferred much of its manufacturing to China to benefit from cheaper labor costs then, as Chinese labor costs continue to rise, China could transfer much of its manufacturing to Africa.

But, again, if labor costs are not going to be all that important, this process may never get going. So perhaps Africa will never be able to enjoy the export boom that led the way to economic growth for so many countries in Asia. In that case it would have to rely on homegrown sources of demand. But, with domestic labor so cheap, it is perfectly plausible to imagine these countries being laggards in the deployment of robots and AI and thereby remaining well behind in the development stakes.

This fear about the development ladder being knocked away from many developing economies has recently taken something of a blow from a study published by the IMF. It argued that essentially there was nothing special about manufacturing. The growth of manufacturing is not a prerequisite for the advance of developing economies nor is it the key to preventing a major gap between "good" and "bad" jobs.

Moreover, the study found that several service sectors exhibited productivity growth equal to the top performing manufacturing industries. It cited postal services and telecommunications, financial intermediation and wholesale and retail distribution. The study concluded that "skipping" a traditional industrialization phase need not be a drag on economy-wide productivity growth for developing countries.[30]

Conclusion

In the earlier discussion about the impact of the AI revolution on income inequality I argued that it is too soon to be sure of the overall effect and we need to keep an open mind. We should be similarly agnostic about the impact of the AI revolution on regional disparities. But we can probably be more confident about the effects of robots and AI on income disparities between countries. These effects will depend partly upon how much countries spend on AI. But this will affect primarily their role in producing robots and AI.

The really important differential factor, though, is going to be how readily countries embrace the *use* of robots and AI or seek to tightly regulate and/or tax them. On these issues, China looks a likely winner in the AI stakes. In that case, of course, since Chinese per capita GDP is well below per capita GDP in America and Europe, the effect of the AI revolution could be to reduce global inequality, much as has been true of globalization over the last two decades.

All these possible effects of the AI revolution on inequality may call for a policy response. Certainly, such action cannot be left to individuals or companies. It must inevitably fall to the state. Indeed, as we prepare for the AI economy, the state potentially needs to be at the center of three major policy issues:

- The regulation and possible taxation of robots and AI.
- Radical reform of the education system to prepare people for both work and leisure in the Robot Age.
- The possible redistribution of income including, perhaps, through the introduction of a universal basic income (UBI).

In regard to these matters, it is high time for us to move from discussion and speculation to action – or at least to the contemplation of it.

PART III

What is to be done?

7
Encourage it, or tax and regulate it?

"Unfortunately, robots do not pay taxes."

Luciano Floridi[1]

"I suppose it is tempting, if the only tool you have is a hammer, to treat every problem as if it were a nail."

Abraham Maslow[2]

Are developments in robotics and AI good for humanity or bad? This question has been at the root of much of the material discussed in previous chapters. It gives rise to a whole series of policy questions. Should we aim to stimulate and encourage developments in AI? If so, how? Or should we aim to restrict them or, at the least, slow them down? And again, if so, how? Is there a good case for a "robot tax"?

Over and above these questions, the preceding chapters have revealed many of the ethical, regulatory, and legal issues that AI gives rise to. Whether AI is good or bad for human happiness, these issues must be addressed. And they can only be addressed by public policy, chosen and implemented by government, in all its various facets.

I begin by asking whether it is advisable, and feasible, to try to inhibit or encourage the use of robots and AI in principle, before discussing the arguments for and against a robot tax. Next, I address regulatory, legal, and ethical issues, before going on to discuss efforts to prevent cybercriminality and cyberterrorism. I conclude with a discussion of the implications of AI for democratic political systems.

Should we deter robots and AI?

The argument for deterrence is that the AI revolution threatens human wellbeing. Accordingly, it is in the public interest to discourage the spread of robots and AI. The displacement of humans by machines supposedly threatens mass unemployment and/or a reduction of workers' incomes and a sharp rise in inequality. For many people the absence of work will bring boredom and alienation as well as poverty, with all the usual social ills that generally accompany that state.

A variant of this argument is that, even if AI and robots are not bad for humanity, society will find it very difficult to adapt to the massive changes involved, and that therefore there is a public interest in delaying them in order to give people and institutions time to adjust.

Yet, as I have argued in this book, there is a strong case that the robotics and AI revolution will do exactly the opposite to these feared outcomes. This revolution is going to enhance our productive capabilities, and in the process afford the possibility of a life of both increased consumption and increased leisure. As I argued in Chapter 5, if the process is left to its own devices, this will probably lead to the destruction of some old jobs, the enhancement of others and the appearance of some completely new ones. In many sectors, human labor and robots will be complementary.

But we cannot know in advance exactly where the destruction/enhancement/creation effects will fall, let alone their magnitudes. Nor can we know the extent to which humans will prefer leisure over work and what they will want to do with whatever extra leisure time they choose. To a considerable extent, this is an area where letting the market find its way will be a discovery process for everyone. It would be potentially very damaging to impose policymakers' choices or presumptions in advance, rather than letting the market do its job.

This is particularly true with regard to the work/leisure choice. Why should the decisions that people make at this critical juncture in human history be seriously constrained by the perceptions of policymakers? They are quite capable of causing great harm even when the thing that they are trying to influence is well established and well understood. But when they are operating in the realm of the unknown, the capacity for harm from mindless intervention is boundless – even when it is well-intentioned.

The AI visionary Kevin Kelly has the right idea. He has written:

> *Many of the jobs that politicians are fighting to keep away from robots are jobs that no one wakes up in the morning really wanting to do.*[3]

Accordingly, I would argue that, although there would be a case for publicly funded programs to assist those made redundant by these, and other, technological changes, as well as assistance in retraining, there is no good case for attempting to discourage or even slow the spread of robots and AI.

Encouragement of AI?

Does this mean that there is a case for governments actively to encourage and stimulate the AI revolution? Andrew Moore, dean of the Computer Science School at Carnegie Mellon University, is adamant. He has said: "Many of us believe there's a moral imperative to get the technologies out there. I find it offensive to deliberately hold back from this."

Mind you, this amounts to an argument for not inhibiting the development of AI rather than an argument for deliberately encouraging it. Indeed, the main argument against encouraging it is effectively the same as the argument against discouraging it. We simply don't know enough about how AI will affect the economy and society. Moreover, the only way to find out is to let people and businesses make free choices. The last thing that government should be doing is actively encouraging AI when heaven knows what distortions and costs this may generate.

Mind you, there is a case for some government action. Jim Al-Khalili, Professor of Physics at Surrey University and President of the British Science Association, who we have met before in this book, worries that unless the government and other official bodies make a serious effort to educate the general public about the risks and benefits of AI, and engage in a full-blown discussion with them, then there could be a serious public backlash against AI, similar to the one a few years ago against genetically modified crops.[4]

Support for a robot tax

I do not think such a course of action makes sense, but if the judgment is made by society that efforts should be made to discourage, or at least slow down, the spread of robots and AI, then one of the most appealing ways of doing this is through the tax system. After all, it would seem as though any taxes levied on robots would be paid by "them." (In practice, of course, all taxes end up being paid by some humans somewhere.)

And the idea of a "robot tax" has garnered considerable support, including from none other than Bill Gates. He argues that "right now, the human worker who does, say, $50,000 worth of work in a factory, that income is taxed and you get income tax, social security tax, all those things. If a robot comes in to do the same thing, you'd think that we'd tax the robot at a similar level."[5] Gates believes that "warehouse work, driving, room cleanup, there's quite a few meaningful job categories that [will be replaced by machines], certainly in the next 20 years."[6]

South Korea has already made a move in the direction of deterring the use of robots. It has announced a limit on the tax incentives applicable to investment in automated machines. In France, the Socialist Party candidate in the 2017 presidential election, Benôit Hamon, campaigned on the idea. (Admittedly, he didn't win. Or even get close.) And, in the UK, the idea of a robot tax has been endorsed by Jeremy Corbyn, the leader of the Labour Party. He told the Labour Party Conference that "We need urgently to face the challenge of automation – robotics that could make so much of contemporary work redundant." He has developed a plan to tax robots and artificial intelligence in order to fund adult education.[7]

Moreover, a robot tax was proposed in the EU Parliament, on the grounds that "levying tax on the work performed by a robot or a fee for using and maintaining a robot should be examined in the context of funding the support and retraining of unemployed workers whose jobs have been reduced or eliminated."[8] The proposal was rejected.

Indeed, Andrus Ansip, the European Commissioner for the Digital Single Market, is strongly against the robot tax, saying that if the EU adopted a robot tax, "somebody else would take this leading position." He thereby opened up the idea of international competitiveness. "Introducing a robot tax in one country when it is not implemented elsewhere might just lead to innovation elsewhere, and both companies and skilled workers moving to a country with a more favorable tax system."[9]

Unsurprisingly, leaders of the technology industry and some international organizations have dubbed the robot tax an "innovation penalty." The International Federation of Robotics believes that introducing the tax "would have had a very negative impact on competitiveness and employment."[10]

The arguments for and against a robot tax

So how do the pros and cons of a robot tax stack up? I have already argued that the general case for restriction and/or discouragement of robots and AI doesn't stack up. Nevertheless, there are three specific fiscal aspects that can be taken to make the case for a robot tax, and they need to be analyzed in turn.

Fiscal neutrality

The first is the argument that, if robots and AI are *not* taxed, then this introduces a damaging distortion into the system. The employment of human labor definitely is taxed, not just through the imposition of income taxes on employees but also through employment taxes, such as National Insurance in the UK and social security taxes in other countries, on both employees and employers. Without a corresponding tax on robots and AI, the argument runs, the tax system is being far from neutral. It is actually encouraging the substitution of robots and AI for human labor.[11]

Accordingly, the tax system could be worsening the problems of technological unemployment, depressed wages, and increased inequality. Even if you do not buy the mass impoverishment case, which I don't, the implication would still be diminished income for society as a result of a distorted allocation of resources.

But what is neutral and not neutral in the tax system depends critically upon whether you regard the robots and AI systems that may replace human workers as "artificial workers" or items of capital investment. If they are seen as artificial workers, then it would be odd to tax them less than human workers. It would be a bit like taxing one sort of worker differently from another – short workers as opposed to tall ones, ones born in the first half of the year differently from those born in the second half, and so on and so forth.

Yet, once you view robots and AI as machines, the complexion of the issue changes radically. After all, not only do we decidedly not put a tax on machines (capital investment) but in many countries we specifically do the opposite through the provision of subsidies, often in the form of various sorts of tax allowance. And yet such capital investment may well result in a reduction in the number of people employed in a particular trade or occupation.

This favorable tax treatment of capital investment rests on three key assumptions:

- Jobs lost in one occupation or sector would always be made up for by jobs created in other occupations or sectors. If this result did not occur naturally through the normal workings of the market, then various government programs to ease structural difficulties, supported by the authorities' monetary and fiscal policies, designed to achieve full employment, would ensure that it nevertheless transpired in the end.
- Society as a whole has a strong interest in achieving high levels of investment, which is the route to higher levels of GDP per capita and hence higher living standards. For a variety of reasons, companies may be

reluctant to invest as much as is in the public interest. Hence the idea of giving some incentive through tax breaks to invest more. This is particularly true where the capital investment involves an element of innovation (as is undoubtedly true with robots and AI). Indeed, because innovations have benefits for society above and beyond the profits that are generated for the innovators, former US Treasury Secretary Lawrence Summers argues that "there is as much a case for subsidizing as taxing types of capital that embody innovation."[12]

- In an increasingly globalized world, if one country taxes capital equipment, this would make it more likely that less capital investment will take place within its borders. Moreover, parts, of all, of the whole activity, firm or industry may transfer to another country.

Given these three assumptions, it is not attractive to equalize the taxation of capital equipment and labor by raising the effective tax rate on the former. And why should robots and AI be regarded differently from other sorts of capital equipment? It would be extremely odd to tax robots and AI but not other sorts of machinery or software that may be equally injurious to the short-term interest of workers whose jobs are threatened by such investment.

In fact, if robots and AI were subject to a special tax, this would lead to some gross distortions. After all, what is a robot? And what is AI? What about bank ATM technology that has undoubtedly eradicated some banking jobs. Does this qualify? Or accounting software programs?

I made clear at the beginning of the book that there is no generally agreed definition of robots and AI and this is for sound reasons. David Poole, Chief Executive of Symphony Ventures, said: "A robot is not a unit equal to a human. Most are not physical robots, they're software robots. It's no different, really, to a spreadsheet." Accordingly, if governments attempted to impose a robot or AI tax, there would be huge legal and tax avoidance and evasion issues, as well as the substantial diversion of investment into "almost robots" and "almost AI."

Perhaps the appropriate takeaway from the debate about taxing robots and AI is not that similar employment taxes should be imposed on them (whatever they are) but rather that such taxes should be abolished for human employees.

But, of course, other things equal, this would leave a hole in the public finances that would have to be filled by increasing some other sort of tax (or by cutting expenditure). Yet people are not queuing up for any sort of tax increase or expenditure reduction. Moreover, as Nobel Laureate Robert Shiller points out, all taxes impose distortions of some sort, so the argument that a robot tax would cause distortions does not satisfactorily close the matter. It is all about how serious the distortions are and what the alternatives are.[13] But the definitional problems of separating out robots and AI from other forms of capital investment, referred to above, surely suggest that the distortions caused by a "robot tax" would be very damaging.

Revenue loss

Nevertheless, there is a second fiscal aspect of the case for a robot tax. In the USA about 80 percent of all federal tax revenue derives from income or payroll taxes and in other countries the equivalent proportion is surely similar. As human labor is replaced by robots and AI, unless similar taxes are imposed on them, the tax base will fall or even, in extreme cases, collapse. So, the argument runs, there is a serious need to plug the revenue gap.

Yet, whether the replacement of human workers by robots and AI leads to a reduction in tax revenue or not depends upon what happens to those human workers. If they become unemployed, then there will be a net loss of revenue – but not if they become employed elsewhere. In that case, the displaced workers will continue to pay taxes, including the wretched employment taxes.

Moreover, the increased output made possible by the use of robots and AI will lead to *increased* revenues to the authorities as people overall pay as much or more income tax as before, but corporations pay more corporation tax and consumers pay more consumer taxes. This has been the experience since the Industrial Revolution. On this basis, there will be no erosion of the tax base and no consequent need to raise extra revenue. Indeed, quite the reverse.

The only major qualification concerns the extent to which workers choose to take the AI dividend in the form of increased leisure rather than increased output (and spending). Since leisure is not directly taxed, this might not lead to an overall rise in tax revenues, and indeed may well lead to a fall. Any fall in revenue would have to be replaced by increasing tax revenues from somewhere or other. A tax on robots and AI is a possible candidate.

In fact, this does not constitute a good case for taxing robots and AI. As I argued in Chapter 4, people are unlikely to take out all of the AI dividend in the form of increased leisure. And if they just take some of it in this form, and the rest in the form of increased income and spending, as I argued is likely, then there is ample scope for the tax base to increase. Moreover, even if it doesn't, for the reasons given earlier, extra revenue needs to be sought in the least distorting ways. As argued above, it is unlikely that a "robot tax" qualifies.

Increased government spending

The third fiscal aspect of the case for a "robot tax" is a further development of this second one. Not only might government revenue fall as humans are replaced by robots and AI but, correspondingly, there may be a large increase in government spending on unemployment benefits and the like. Indeed, it may well be that the mass disappearance of jobs makes it desirable to introduce some form of universal basic income (UBI) or guaranteed minimum income (GMI). (I will review the arguments for and against this suggestion in Chapter 9.) In this case, so it is argued, it is both right and efficient to tax the very thing that is responsible for this increase in government spending, namely the spread of robots and AI.

Of course there may be particular individuals, or groups, that are severely disadvantaged by the spread of robots and AI. In extreme cases they may lose their livelihoods as their skills are made redundant. They will have to seek reemployment elsewhere, and this may involve the costly acquisition of new skills. Some of these people may be incapable of acquiring the skills that are in demand in the new world. Accordingly, there is a strong case for public funding to ease the transition from one form of employment to another and to sustain those who, for one reason or another, cannot find employment at all.

But if it is a matter of revenue needing to be raised to provide finance for measures to ease the transition or to alleviate hardship, or to finance the UBI then, as with any other sorts of government expenditure, the revenue should be raised by the least distorting methods across the whole tax base, in all the usual ways. Particularly given the definitional issues discussed above, it seems unlikely that the best – that is, least distorting – way of raising revenue is by levying a tax on robots and AI.

So the issue of taxing AI in order to slow its spread and hence protect jobs for humans needs to be separated from the question of whether some form of increased public assistance, such as a UBI, should be introduced in order both to cushion the ill-effects of AI's development on particular individuals and groups and to provide a way, more generally, of reducing the disparity of incomes.

Robots and public policy

Just because the robot tax is a bad idea and allowing the market to decide on the appropriate spread of robots and AI is a good one, this does not mean that government can simply stand back from the whole AI field in a spirit of complete laissez-faire. There are a series of issues concerning robots and AI where the involvement of governments is essential.

Many years ago, the visionary Isaac Asimov[14] recognized the potential for robots to do harm and hence the need for an ethical (and perhaps legal) framework governing their use and development. He laid down three principles which, although they leave many specific issues unaddressed, constitute a good starting point for thinking about both private conduct and public policy:

- A robot may not injure a human being or, through inaction, allow a human being to come to harm.

- A robot must obey the orders given it by human beings except where such orders would conflict with the First Law.
- A robot must protect its own existence as long as such protection doesn't conflict with the First or Second Laws.

The European Parliament recently voted for the creation of an ethical–legal framework in relation to robots. Surely, this is quite right in principle. But heaven knows what the EU will make of such a framework in practice. As with all sorts of regulation, there is a danger that regulatory measures could be used effectively to deter innovation as a way of protecting particular sorts of employment, or employment in general. This would be fully in accordance with the history of economic and commercial development, and fully in accordance with both the history and the innate tendencies of the EU. So, it is important for governments to get the balance right.

Restrictions on AI research

Even if we do not try to restrict or discourage the employment of robots and AI, there could be an argument for restricting AI *research* in order to prevent future developments that are against the public interest.

At the outset, I should say that a general restriction of research only begins to make any sense if you adopt an ultra-pessimistic view of the implications of robots and AI for humanity. Perhaps if you embraced some of the radical views about humanity's fate that I discuss in the Epilogue, this might make sense. But on anything like the view of the implications of robots and AI for humanity's future expounded in this book, any attempt to restrict or discourage general AI research would be absurd.

But there are some exceptions and limitations to this conclusion. Where certain types of development in robotics or AI can be shown to be harmful to humans, not in the general sense, discussed above, but in some specific respect, then there could be a case for the introduction of restrictions on research into those applications.

At a micro level, some restrictive action is already taking place. The South Korean research university KAIST has just encountered a boycott from more than 50 of the world's leading robotics experts over its decision to open an AI weapons laboratory. Several countries, including the USA, Russia, China, Israel, and South Korea are developing autonomous weapons that can choose courses of action and implement them without intervention from a human controller. The UN and Human Rights Watch have advocated a ban on such weapons.

Yet whether more general and widespread restrictions are feasible is another matter. Attempts to prevent or even slow scientific or technical progress do not have a successful history. Let us take nuclear weapons as an example. It is true that there has been no nuclear exchange since two bombs were dropped on Japan at the end of the Second World War. Although the number of countries possessing nuclear weapons has increased, to some extent it has been kept in check by treaty. Similarly, the use of chemical and biological weapons has also been restricted by international treaty, with considerable (but not total) success. But all this is about use. There is no evidence to suggest that the human race has stopped accumulating knowledge about how to manufacture or deliver such weapons of mass destruction or increase their destructive power.

By contrast, as far as we know, effective research into eugenics and experimentation on human beings of the sort that the Nazis carried out has been halted, although primarily by domestically-based ethical and legal restraints rather than by international treaties. Yet this is a very different issue, not least because of the demonstrable harm inflicted on large numbers of individuals and the recentness of the Nazi experience and its aftermath.

It is going to be particularly difficult to restrict AI research when the immediate prospect is of so many benefits to human beings and when the possibly dark prospects beyond the Singularity apparently lie so far off. Moreover, we again run up against the definitional issue. If you are going to try to ban or restrict research into robotics and AI, how would you define the boundaries between these and other sorts of machinery and software?

In addition, there is the age-old problem of common action. One country, going it alone, could decide that the dangers facing humanity are so great that it should effectively repress AI research and perhaps even seek to impede the employment of robots and AI in society. But if one country decides to go down this route alone, it will almost certainly fall down the international pecking order. That has severe implications for the ability of that country to defend itself and/or to influence world affairs. This points to the need for international agreement on restricting AI research. Yet it is extremely difficult to stop the onward march of knowledge in any one country, let alone to get international agreement on it.

And if the whole world was not signed up to restraint, things could get very nasty. Consider what would happen if the rest of the world restricted such research but one rogue government did not. The latter might eventually have the capacity to completely dominate the rest. And consider what would happen if ISIS, or some other such organization, continued to develop AI while the rest of the world stood still. It does not bear thinking about.

Criminal applications

One aspect of AI that calls for serious state intervention is the prevention of cybercrime. This has many possible manifestations. One is the development of advanced malware in order to steal browsing history or personal information with the intention of using it to commit fraud or hack accounts. (Malware includes viruses, spyware, and other unwanted software that gets installed on a computer or mobile device without consent.) It has been reported that the number of different malware jumped from 275 million in 2014 to 357 million in 2016.[15] Moreover, machine-learning systems that recognize patterns in data and are able to "learn" without being specifically programed, introduce further risk.

An AI program could be created to automatically monitor personal emails or text messages and to create personalized "phishing mail." This involves the use of a scam email or text message alerting someone of a security breach, encouraging them to click on a link and give away their personal information. With the help of AI these scam messages become more personalized, and consequently the security systems will require greater intelligence to identify them.

The use of AI could lead to what security software specialist Symantec describes as, a "full-fledged arms race between cybercriminals and security." As George P. Dvorsky has put it: "with AI now part of the modern hacker's toolkit, defenders are having to come up with novel ways of defending vulnerable systems."[16] But, as in so many other instances, there is a serious question about the cost of measures that would effectively deal with the problem. If these are exorbitant, the better course of action may be simply to live with it.

AI and terrorism

Terrorists are also starting to use AI for encryption purposes. Extremist groups have started to employ "virtual planner" models of terrorism, as a way of managing lone attackers. Top operatives are able to recruit members, coordinate the target and timing of attacks, and provide assistance on topics like bomb-making, without being detected. Furthermore, there are fears that terrorists will acquire autonomous vehicles and drones to perform attacks.

Naturally, just as in conventional warfare, these dangers have led to a corresponding increase in defensive activities. Facebook has counterterrorism detection systems that operate using AI. It says that it removed or flagged 1.9 million pieces of content linked to terrorism in the first part of 2018, nearly a twofold increase over the previous quarter. Nevertheless, its defensive moves may still be inadequate. In March 2018, the European Commission published guidelines increasing pressure on social media platforms such as Facebook and Twitter to detect and delete terrorist or radical material within an hour of it being uploaded.

There is surely still massive room for improvement in society's ability to resist damaging activity by criminals and terrorists. Although the Electronic Communications Privacy Act, passed in 1986, allows the US government to access digital communications such as email, social media messages, and information on public cloud databases, encrypted messages are still able to go undetected.

There have been proposals for governments to have access to privately encrypted messages, giving rise to a heated discussion. A group of academic computer scientists has concluded that these proposals are unworkable in practice, raise enormous legal and ethical questions, and would undo progress on security at a time when internet vulnerabilities are causing extreme economic harm. And if the government had access to every person's private messages, then a hacker would need only to break through the government's wall to gain mass access to all this information.

Legal and insurance issues

Even without any attempt to restrict or discourage developments in robotics and AI, there needs to be a regulatory and legal framework governing their operation. The first issue concerns the ability of a robot or AI to cause damage or inflict harm on humans. There is a key question about where liability for any such damage or harm lies. Does it lie with the robot's user, with the robot's manufacturer, or with its designer? Or are none of the above to be held responsible? Are all members of the public interacting with robots and AI to be responsible for their own fates?

Roman law supplies a possible answer in the form of the law in relation to slaves. If a slave caused an accident or inflicted damage on someone else's property, the responsibility rested with the slave's owner. As regards legal liability, could we perhaps treat robots and AI as though they were slaves under Roman law? (Surely not once the Singularity has occurred. But that must wait until the Epilogue.)

Take, for instance, our well-worn example of driverless cars. When these vehicles are involved in accidents who is to be held responsible? What obligations fall upon other, human, drivers when interacting with driverless vehicles? What are the implications for insurance companies? Such problems will require careful drafting of the relevant laws.

This is not a mere optional extra. Without legal clarity there will be a substantial inhibition to employing AI. This could lead to a considerable gap developing between what is technically feasible with AI and how much it is employed in practice.

This task can only be undertaken by government. Indeed, if there is one thing that government can do to encourage the spread and take-up of AI it is to bring about clarity with regard to legal and insurance liability. The Royal College of Radiologists told the House of Lords Select Committee on Artificial Intelligence that "legal liability is often stated as a major societal hurdle to overcome before widespread adoption of AI becomes a reality."[17]

But there is also scope for private action to ensure safety and promote public acceptance of AI. In June 2016 Google DeepMind was reported to be working, alongside some Oxford academics, on developing a "kill switch," that is to say a code that would ensure that an AI system could "be repeatedly and safely interrupted by human overseers without [the system] learning how to avoid or manipulate these interventions."[18]

In order both to ensure safety and to bolster public confidence transparency is essential. For instance, patients are unlikely to be prepared to accept a treatment that is the outcome of some algorithm's analysis of the data. They are likely to want a clear justification from a qualified expert. But for that expert's view to be worth anything then the route to the algorithm's conclusion must be capable of being understood by that person.

Professor Alan Winfield of the Bristol Robotics Laboratory told the House of Commons Science and Technology Committee that it was important to be able to "inspect algorithms so that if an AI system made a decision that [turned] out to be disastrously wrong ... the logic by which the decision was made could be investigated."[19]

Microsoft's Dave Coplin told this same Committee that "in AI every time an algorithm is written, embedded within it will be all the biases that exist in the humans who created it." He stressed the need "to be mindful of the philosophies, morals, and ethics of the organizations ... creating the algorithms that increasingly we rely on every day."

Interestingly, there is now a growing recognition of the need to regulate humans' interaction with robots and AI, not because of their capacity to do harm but rather because some people and institutions, notably the police and the civil service, may try to blame robots and AI for their own mistakes. In September 2018 a report from the Royal United Services Institute noted that in relation to the increasing use of algorithms to decide whether a suspect should be kept in custody, there was a lack of "clear guidance and codes of practice."

The power of data

Further legal and ethical issues arise concerning data. Many of the useful applications of AI derive from machine learning and that depends upon the processing, analysis, and manipulation of huge amounts of data. Hence the expression "Big Data." This data involves the preferences, habits, behaviors, beliefs, and connections of individuals on a massive scale.[20] The collection and analysis of this data provides a source of knowledge about individuals that may well improve the supply of goods and services that are marketed to them. But this may readily invade their privacy and compromise their rights.

When aggregated and analyzed, this data provides a source of knowledge about aggregate behavior that can be invaluable to some companies trying to fine-tune their offering to the market. But it can also lead, among other things, to the possibility of influencing and manipulating elections. And running through this whole issue is the question of who "owns" the data and has the ability and legal right to pass it on to other potential users, with heaven knows what consequences.

There is here a legitimate case for state action. As Zeynep Tufekci, a University of North Carolina Professor who studies the social implications of technology, puts it: "privacy of data simply cannot be negotiated person by person, especially because there's no meaningful informed consent. People cannot comprehend what their data will reveal, especially in conjunction with other data. Even companies do not know this, so they cannot inform anyone."[21]

The resale of user data is a major problem area. In an article for the *E-Commerce Times*, Pam Baker states that "data monetization is Facebook's business model. Facebook and some other tech firms exist solely to gather and sell everyone's data, exposing users' lives in increasingly more granular detail."[22]

There was a recent scandal involving the company Cambridge Analytica, which took data from Facebook and allegedly used it to try to influence voter opinion. A Facebook app, produced by a Cambridge

University academic outside his work at Cambridge, collected data on around 87 million Facebook users.[23] The company was then able to develop "sophisticated psychological profiling and personalization algorithms."[24] Politicians could then hire the company to try to influence voter opinion. Cambridge Analytica's managing director told an undercover reporter that "it's no good fighting an election campaign on the facts." Instead, they believe that "actually it's all about emotion."[25]

Fear of what might happen to data about them is starting to influence people's behavior. Research conducted by Mintel, a global market research firm, found that 71 percent of British consumers actively avoid creating new accounts with companies. In fact, the British are most concerned by the safety of their financial data, with 87 percent claiming that they are concerned about sharing these details with companies. As a result, the company's senior technology analyst, Adrian Reynolds, said that "the increasing use of connected devices to access websites and apps is producing a wealth of personal data sharing, making it very difficult for consumers to keep track. For many, limiting further exposure is their preferred option."[26]

Although there are only limited policies and laws in place to do with AI specifically, there are already measures in place aiming to protect personal data. The EU's General Data Protection Regulation (GDPR) was introduced in May 2018, "designed to protect and empower all EU citizens data privacy and to reshape the way organizations across the region approach data privacy." Under the GDPR, only the data deemed absolutely necessary will be held by companies, and in the event of a data breach companies are required to inform the relevant authorities and their customers within 72 hours. Failing to comply with the GDPR could result in organizations being fined up to 4 percent of annual global turnover.[27]

Human freedom and dignity

AI presents worrying threats to human liberty and privacy. A network of AI-enabled surveillance cameras, along with the monitoring of internet activity, could be employed by governments to engage in mass-surveillance measures.

For example, the EU provides funding for a research project called INDECT, conducted by several European universities. The project encourages European scientists and researchers to develop solutions and tools for automatic threat detection. Some have accused INDECT of privacy abuse. In an article for *The Telegraph*, Ian Johnston called it an "'Orwellian' artificial intelligence plan." Shami Chakrabarti, the former director of human rights group Liberty, described the introduction of such mass-surveillance techniques as a "sinister step" for any country, and that on a European scale it is "positively chilling."[28]

Not only are the EU, UK, and USA showing an interest in aspects of mass surveillance, but Asia is investing heavily in it. Chinmayi Arun, Research Director of the Centre for Communication Governance at NLU Delhi, speaks about the threats that AI poses to civil liberties and democracy in India. She argues that "in democracies like ours, the balance of power between the citizen and the state is really delicate, and there is a great potential for AI to tip that balance of power in favor of the state."[29]

In China, this idea of mass surveillance has resulted in the government's 2020 project, which states that a national video surveillance network will be "omnipresent, fully networked, always working and fully controllable."[30] A similar project is already in place in China's Xinjiang region, where facial recognition, license plate scanners, iris scanners, and CCTV create a "total surveillance state."[31]

What's more, the 2020 project will launch the Social Credit System nationwide, with a mission to "raise the awareness of integrity and the level of trustworthiness of Chinese society."[32] Mass surveillance of CCTV and social media record a person's behavior, and, as a result, each citizen will have their own "Citizen Score." If a person's score becomes too low, they can be punished by being unable to purchase travel tickets or make reservations at restaurants. Several provinces have actually been using TVs or LED screens in public places to humiliate and expose people, and some have personalized dial tones of blacklisted debtors so that callers will hear a message akin to "The person you are calling is a dishonest debtor."[33]

Rogier Creemers, who studies Chinese governance at Leiden University, highlights why China's use of mass surveillance is more extreme than other countries. He states that the difference in China is the historical context: "Liberal democratic institutions are based on the notion that state power must lie in the hands of the population. There are things the state is just not supposed to know or do." But, he says, "China starts from a different point of view – that a strong empowered state is necessary, in order to drag the nation forward. In China, surveillance is almost a logical extension of what the state is supposed to do, because the state is supposed to keep people safe."[34]

Concerns about personal liberty should not be restricted to the activities of the state. Private entities should also be a source of concern. In China, workers are being equipped with hats and helmets with electronic sensors fitted. These enable employers to read employees' emotions.[35] One can readily imagine what they will discover, and shudder at what the implications may be.

On a completely different note, do we humans have to be concerned with the rights and welfare of robots and AIs? Some people have started to think about such questions. While robots and AI are mindless automata, they cannot think or feel and any "decisions" they may take will ultimately derive from their human originators and owners. This is the current state of play. So, as things are now, robots and AI do not raise any particular moral or ethical issues.

Of course, things will not necessarily stay this way. Once robots and AI can think, feel, and make independent decisions, then some serious ethical issues may arise. But we can defer examination of such tricky matters until the Epilogue, when the Singularity will be almost upon us.

What AI means for political structure

Of much more immediate relevance is the potential impact on democracy. Many AI experts have concluded that advances in robotics and AI will transform the shape of society as an elite few lead lives far removed from the "lumpen leisuriat." The structure of society would come to resemble how things looked during the Middle Ages, except that then the serfs worked. Indeed, without their work, the lords would have had no wealth or income. In the new world, by contrast, the masses would not work – and this would not matter for the wealth and welfare of the elites.

Who would own and control the armies of robots and AIs which would be the source of all wealth and power? Many AI visionaries suggest that this is bound to lead to the end of democracy. The future, they say, lies with some sort of dictatorship or, at best, an oligarchical form of government.

As Nigel Shadbolt and Roger Hampson put it: "… the problem is not that machines might wrest control of our lives from the elites. The problem is that most of us might never be able to wrest control of the machines from the people who occupy the command posts."[36]

This would be ironic because the World Wide Web and much subsequent digital development originated in a spirit of extreme libertarianism. Sir Tim Berners-Lee, the founder of the internet, did not establish any patents or property rights in his creation because he wanted it to be free. The wired world is supposed to be a levelling force in society and its effect is supposedly anti-authoritarian and anti-hierarchical. At least in the West the internet does not belong to governments. Mind you, this is evidently not true in China.

But, as we have just discussed, even in the West there are serious threats to individual freedom and rights to privacy, even from within the private sector. Moreover, the creators of the great digital-era companies that dominate the modern world – Amazon, Google, Apple, and so on – are decidedly of a kind: the Western ones are all American, white, and male. Typically, although the founders have become mega-rich by selling stakes in what they have created, they retain control.

The 2012 presidential campaign by Barack Obama used machine learning and big data. As a result, the campaign was "highly successful in not only mobilizing, but also convincing voters to give Obama their support."[37] Then, during the 2016 US presidential election, the firm Cambridge Analytica used big data and machine learning to send voters different messages based on predictions about their susceptibility to different arguments.[38]

Another issue that has emerged recently is the use of fake news to influence election campaigns, arguably preventing a "fair" election. While "fake news" has to some extent always existed, social media and machine learning "allow fake news to be spread more deliberately and effectively."[39] Bots are often effective at targeting supporters of opposing parties and discouraging them from voting.

Bots were used days before the most recent French presidential election to popularize "#MacronLeaks" on social media. Facebook and Twitter were inundated with reports building a narrative of Macron as a fraud and hypocrite.[40] Mind you, this didn't stop him from being elected.

There has been some debate over who should be held responsible for the publication of fake news. Companies such as Facebook and YouTube have claimed that they are just a "platform" rather than a "publisher," and that they are not responsible for posted content.[41] However, Facebook did admit that it was too slow in stopping people using the social network to "corrode democracy" during the US election.[42] French President Emmanuel Macron told *Wired* magazine that, if left unchecked, AI could "totally jeopardize democracy."[43]

There is a wider politico-economic matter that feeds into this issue. Does the world of AI and robots naturally have an antidemocratic tendency, associated with what it does to the distribution of income? If so, can the redistribution of income to achieve a more egalitarian outcome fend off this result and preserve democracy? Or would it, perhaps, intensify the problem? I will address this question in Chapter 9.

Of course, this isn't necessarily the problem that it is cracked up to be by the AI pessimists. As I have argued in previous chapters, it is far from a done deal that robots and AI will bring on mass unemployment and/or impoverishment. Indeed, I have argued just the opposite.

The need for public action

The outcome of this chapter amounts to neither a case for radical restrictions nor an argument for complete laissez-faire. As regards the need for public action, the conclusions are a mixed bag:

- There is no compelling case for a tax on robots and AI. Moreover, the practical difficulties facing such a measure are enormous.

- Nor is there a compelling case for restricting general research into robotics and AI. In any case, an attempt to do this would be likely to founder on practical difficulties.
- Nevertheless, there is a case for restricting research where its applications are clearly harmful, including where they would further the activities of criminals and terrorists. Again, though, the practical difficulties will be huge. A more promising avenue with regard to countering cybercriminality and cyberterrorism is probably to use public funds to support research into how such activities can be countered with the help of robots and AI.
- But even if it takes no action to promote or restrict robots and AI, the state needs to take action to develop a far-reaching legal and regulatory framework governing them. Indeed, without this it will be difficult for businesses to make full use of the opportunities that they provide.
- There is also an urgent need for the further development of the framework governing the use of data.
- By contrast, concerns about the welfare of robots and AI can wait until the Singularity is almost upon us – if that ever happens.
- But someone, presumably not government, urgently needs to study the implications of the new AI world for political structures and institutions.

Perhaps this last item should be one of the key assigned essay tasks for all those wishing to be educated for work and life in the Robot Age – that is, of course, if the art of essay writing survives the AI onslaught. For, make no mistake, some of the greatest implications of the AI revolution lie in the field of education. But they are not necessarily what you imagine.

8
How the young should be educated

> "Education is what society does to you, learning is what you do for yourself."
>
> *Joi Ito* [1]

> "Computers are useless. They can only give you answers."
>
> *Pablo Picasso* [2]

It is tempting to believe that the fundamentals of a good education are unchanging. It is tempting but it isn't true. In medieval Europe the role of universities was to train ministers of the Church, lawyers, and teachers. The subjects they studied reflected those requirements.

Throughout the nineteenth century, and carrying on until the Second World War, the great English "public" schools saw one of their primary roles as to train those who would administer the Empire. During the early part of this period a high proportion of students at Oxford and Cambridge still studied theology and classics. It wasn't until much later that scientific and technical subjects such as chemistry, physics, biology, and engineering came to be widely studied. (And dubious subjects such as economics, politics and sociology came later still.)

That was then. Of course, things now are radically different. Today the study of theology and classics is a minority occupation. Accordingly, we may like to think that the world of education has now "caught up" with modern reality and so now there is no need for further radical change. But actually, this isn't true either. The education system is one of the most antiquated and out of sync parts of today's society. And now both the economy and society are about to undergo another fundamental change because of robots and AI. Education must change, too.

There are four main issues that need to be discussed here:

- In view of both changing employment prospects and changing opportunities for leisure brought about by robots and AI, what subjects should be taught at school and university?
- What do robots and AI imply for the *methods* of education?
- What do they imply for the number of years to be spent in full-time education?
- What is the role of the state in fashioning an appropriate and effective education system for the new world?

Education in the Robot Age

Let us begin by thinking about education for the world of work. Ever since AI began as a scientific discipline in the 1950s, it has been an abstruse and even elite subject restricted mostly to graduate students in computer science at major universities.[3] Now, surely, it must enter the mainstream. And, more than that, it must have a key bearing on what children are taught at school. But what bearing?

It is often argued that the future must lie exclusively with STEM subjects – that is, science, technology, engineering, and mathematics. You can see why. In particular, as I explained in Chapter 5, not only will there be many jobs directly connected with robots and AI, but just about everyone will need to cope with them in their daily life. At the very least, people must learn to interact with them as they had to learn to interact with telephones, cars, and computers.

Accordingly, many argue that all those humanities subjects like languages, history, and geography, as well as art, music, and drama, should take a backseat, if not be eliminated altogether. You could say that they are the theology and classics of the modern era. A few students may still study these subjects, probably at graduate level, but they should cease to be part of the core curriculum. All those students who used to flee at the sight of a Bunsen burner or quail at the thought of a quadratic equation, and sought refuge in French irregular verbs, or the kings and queens of England, need to wake up and smell the coffee. (You know who you are.) The nerds are in the ascendant now.

If this is how things are going to have to be, there is a long way to go. In 2016 only about 40 percent of US schools taught computer programing. Only 58,000 students took the AP Computer Science A Exam, compared to 308,000 students who took the calculus test. Although this figure is on the rise, there is evidently still a long way to go until computer skills are regarded as important as general mathematics.

But it isn't only about the percentage of students taking a course whose name suggests relevance to the changing world. ICT skills currently being taught in schools are largely out of date. A more up-to-date framework would incorporate AI into schools' lesson programs. Although in the UK in 2015 the decision was made to replace the ICT GCSE with a revamped computer science qualification, there has only been a small rise in uptake from students.[4] Much of the content now focuses on coding and programing.

When the subject is not compulsory, however, if the interest is not there on the part of the students, then revamping courses might have no effect. IBM developer Dale Lane, who helped create the educational tool Machine Learning for Kids, says "as a result of AI not being on the core curriculum, finding time to dedicate to it in the classroom is a challenge and this is partly why adoption on the ground has been slow."[5]

A severe shortage of teachers trained in computer science is also a major hurdle. In fact, despite the lack of obvious incentives to provide a public good, parts of the private sector are making a serious contribution to driving forward AI education. One possible route to secure more trained teachers could be through collaboration with tech companies. The TEALS program, supported by Microsoft, helps high schools in the USA develop computer science classes.[6] Computer professionals are paired with high school teachers for a few hours a week, in the hope that it will create a ripple effect.[7] A collaboration between educators, governments, and industry professionals may be the way forward.

As it happens, a growing number of primary and secondary schools are teaching pupils how to code. But this doesn't exactly equip students well for the AI economy. Professor Rose Luckin of University College London's Institute of Education believes that this skill will actually be "old hat" by the time that current students enter the workforce.[8]

It certainly isn't just coding that needs to be taught in schools. As the educationalist Ben Williamson says:

> *knowing about privacy and data protection, how news circulates, understanding cyberattacks, bots and hacking, how algorithms and automation are changing the future of work, and that there are programers, business plans, political agendas and interest groups behind this, is worth including in a meaningful computing education too.*[9]

Moreover, it is important that ethics features heavily in any AI-infused educational program. For AI technologies face many serious ethical dilemmas. It is important that those directly involved in developing and operating AI techniques not only are familiar with these ethical issues but also are able to contribute to the debate in society on those issues and be able to interact effectively with policymakers.

In defense of traditional education

So there needs to be more attention paid to AI in the school curriculum. It does not follow, however, that education should be only, or even mainly, about robotics and programing. After all, over the last 50 years, the motor car has been a central part of our life. But hardly anybody has known about the internal workings of their vehicles. Nor have they needed to. Still less have they needed to be taught about their workings at school or university.

Actually, the same goes for computers. Indeed, although the basics of how to operate a computer have been widely taught in schools for many years now, I am not convinced that this has been central to the widespread acceptance and use of computers. Quite apart from students, the rest of society has also had to learn how to use them and this skill has been largely self-taught. (Admittedly, though, there are quite a few older members of society who still don't know how to use them at all, and many more who do not know how to make use of all their capabilities.)

More recently, the smartphone has become ubiquitous. To the best of my knowledge, no schools have taught smartphone usage. Nor have they needed to. People have just picked up the skills as they have gone along. I reckon that interacting with robots and the various forms of AI may be just like learning to use a smartphone.

Nor will the skill of how to interact with robots and AI, whether taught or self-taught, be the dominant requirement, let alone the only one. I have argued in preceding chapters that the AI economy will lead to the full discovery of the human realm. Accordingly, it would be odd for education to be all about the discovery of the machine realm.

So traditional subjects and the traditional approach to education will still have their place. Everyone will benefit from a full, rounded education. And the senior members of society, whether managing businesses, countries or cultural activities, will still need to develop and hone critical thinking skills. This backs up the need for a traditional education, including the study of history, religion, art, drama, music, philosophy, and other cultures.

This conclusion is now starting to be generally accepted by both entrepreneurs and educationalists, who are engaged in an active debate on these issues in both business and education journals. Indeed, a mini-industry has sprung up dealing with the subject. I will give you a flavor of what some of the leading protagonists are saying.

Mark Cuban, the American billionaire, went as far as predicting that, as automation becomes the norm, it will be "free thinkers who excel in liberal arts" who will be demanded.[10] Accordingly, some commentators have argued that arts need to be on a par with STEM subjects. They have suggested another acronym to reflect this – STEAM.

Relatedly, the futurist and technology guru Gerd Leonhard has argued that to counterbalance STEM subjects there is a need to emphasize subjects with a definite human focus. Because we seem to be in the midst of a battle of the acronyms, he has suggested the merits of CORE – Creativity/Compassion, Originality, Reciprocity/Responsibility, and Empathy.[11]

A similar view is taken by Carnegie Mellon Professor David Kosbie and coauthors in an article published in the *Harvard Business Review*. They argue that "with AI taking over routine information and manual tasks in the workplace, we need additional emphasis on qualities that differentiate human workers from AI – creativity, adaptability and interpersonal skills."[12]

The educationalist and thinker Graham Brown-Martin says that we should encourage the development of skills that cannot be replicated by robots. He says that "this is great news because it means we can automate the work and humanize the jobs."[13]

The President of Northeastern University in the USA, Joseph Aoun, has given a name to the new discipline that people will need to learn in the digital age – thankfully without coining another acronym. He calls it "humanics." He says that students will need to build on the old literacies by adding three more – data literacy, technological literacy, and human literacy.[14]

The need for reform

Mind you, even if the continued importance of traditional subjects is accepted, including the arts and humanities, this certainly does not mean that education can stand still. In fact, even before any account is taken of robots and AI, it seems that the qualities that employers most value in employees are not ones that are taught or even encouraged by traditional academic study. According to a 2006 survey, the qualities that employers most value in prospective employees are "leadership" and "ability to work in a team." After that, written communication and problem solving come next. Technical skills came out in the middle of the survey responses – below strong work ethic and initiative.

Admittedly, you could argue that the skills revealed as most valued in this survey are fostered more by a traditional liberal arts education than by a technical one. But is even a traditional liberal arts education particularly good at fostering such qualities? Probably not. If what we want is young people imbued with creativity, initiative, leadership, and the ability to work well in a team, then it is surely right to be seriously critical of the whole modern approach to education. The educationalist Sir Ken Robinson has said that the whole business of penalizing "wrong answers" stifles creativity: "We don't grow into creativity. We grow out of it, or rather we get educated out of it."[15]

Education is sometimes equated with "learning," and in the past considerable stress was put on learning things by rote. There is still a good deal of this today – everything from times tables to the dates of kings and queens. In parts of Asia, rote learning is especially emphasized. Yet surely a good deal of the worth of such learning derived originally from the difficulty (and cost) of retrieving information and knowledge. Better to commit it to memory.

But just about everything about just about anything is now only a few clicks away. Accordingly, you would think that education should rather focus on where to find information, whether to trust it, and how to weigh it. This alone should justify a shift in what education is fundamentally about.

Admittedly, we must beware of throwing out the baby with the bathwater. If we do not try to "learn" any facts at all will we be able to have any understanding? Can we really begin to think about the big issues in history, for instance, without having at least a passing acquaintance with what happened when, what had happened before, and what came next?

That said, as the influence of AI on education grows, we should not simply seek to shore up the existing system of education. The changes afoot because of AI offer an opportunity to rethink fundamentally how the educational system works and what education is for.

The answer that both ancient philosophers and modern educationalists would give to the second question is surely "to improve the lives of those being educated." I, too, would endorse this objective. And, over and above this, there is the objective of improving everybody else's lives, too. For education may give rise to what economists call externalities. In other words, it can have spill-over effects on other people beyond the individuals being educated. This is because better-educated people can take a fuller and more valuable part as citizens, as well as probably adding to the overall productive capacity of the economy, above and beyond the extra income that accrues to the individuals being educated. Some might argue that better education also fosters kindness, honesty, and peacefulness, all of which would have broader benefits for society as a whole. Mind you, this last set of benefits is highly questionable.

The leisure imperative

Preparing for the world of work and participation in society are not the only purposes of education. Preparing people for leisure should also play a part. I argued in Chapter 4 that in the AI economy human beings will probably choose to spend more of their time at leisure. Much of this extra leisure time will undoubtedly be used to do more of the usual activities that people undertake on their own, or with their family and friends. And many people do not need any help in either making full and enjoyable use of this time or in developing new hobbies and activities.

But quite a few people already find it difficult to make full and effective use of their free time. As leisure time increases it will be a challenge for people to use these extra hours in a fulfilling way. Education has a major role to play. It can teach people how to be fulfilled on their own. This may involve teaching them about literature and music or many of the other things that can give enjoyment.

This message can easily sound elitist and patronizing – dispensing the appreciation of Beethoven and Balzac to the masses as "approved" ways of spending leisure time. It seems all too reminiscent of "Big Brother." In practice, of course, educators have no control over how people spend their time. Nor should they.

Yet they can have a decided influence on what people are exposed to during their youth, and this can have beneficial effects for the rest of their lives. At its best, the educational system introduces people to things that they would not ordinarily encounter without its intervention. For many students this will surely include the likes of Beethoven and Balzac. And if that sounds elitist, then tough.

At the very least, encounters with the likes of these two gentlemen, among others, facilitated by the educational system, provide a yardstick against which to measure such stalwarts of popular entertainment as *Celebrity Big Brother* or *Keeping up with the Kardashians.* (Anyone who does not know what these programs are should count themselves blessed. They are specimens of the genus known as "reality TV," in which people are filmed doing anything and everything and uttering various banalities for the entertainment of the people watching at home.)

The notion of introducing young people to things goes right to the heart of what education should be about. The word "education" appears to derive from two Latin roots, the verb *educare*, meaning to train or mold, and the verb *educere*, meaning "to lead out from." And what young people need to be led out from is not only a world of poverty, prejudice, and ignorance, but also the world of *Celebrity Big Brother* and *Keeping up with the Kardashians.* If, having been led out from this miserable, stunted world, they decide to go back in, then that is their lookout. But not to help young people to escape from it in the first place would be a dereliction of duty on the part of those charged with *educating*.

Yet education for leisure needn't only be about things intellectual and cultural. It should also introduce young people to crafts such as woodwork, painting, and cooking. And it should introduce them to sport as something to be played rather than just watched. It can help to develop in them an enjoyment of being physically fit and an understanding of how best to maintain fitness of these things too, large numbers of people in modern society can, without intervention, grow up being largely ignorant, or at least inclined toward inactivity.

In addition, education has a key role to play in teaching people social skills, as well as inculcating the values of community and the responsibilities of citizenship. And it can introduce people to the merits and strengths of the voluntary sector.

All this is a far cry from the thrust of so much recent commentary on the education system that has emphasized the need for education to be heavily academic, with STEM subjects dominant in order best to prepare students for the AI economy, and/or focused on the acquisition of skills or a narrow base of knowledge that leads directly to the prospect of a job. There is far more to life than work. Indeed, there is far more to work than work.

Educational methods

Over and above what subjects need to be taught at school and university in the Robot Age, the methods of education certainly need to undergo a revolution. What is the point, in this day and age, of teachers and lecturers standing before a class or lecture group and reading out lecture notes, which the class/lecture group dutifully copies out? It is astonishing and scandalous that something like this still goes on in most of our educational institutions.

Never mind the advance of robotics, according to the distinguished educationalist Sir Anthony Seldon, Vice-Chancellor of Buckingham University, much of the educational establishment is still employing methods used back in 1600.[16] Actually, I reckon that there hasn't been that much change since the time of Aristotle in the fourth century BCE. Properly used, AI can effectively put an end to all this.

Most people seem to think that the impact of the AI revolution on education will involve a major reduction in the demand for teachers. As Martin Ford puts it:

> *Now imagine a future where university students can attend free online courses taught by Harvard or Oxford professors and subsequently receive a credential that would be acceptable to employers or graduate courses. Who, then, would be willing to go into debt in order to pay the tuition at a third- or fourth-tier institution.*[17]

Actually, such a revolution is already underway. More people have registered for Harvard's online courses than attended the "physical" university in its nearly 400 years of existence.[18]

But I don't think that the conclusion that there will be a massive fall in the demand for school and university teachers is right at all. In any case, the amount of strain currently placed upon school teachers makes the present system unsustainable with the existing resources. In the existing system, either there will have to be more teachers, or some way will have to be found to reduce the demands placed upon them. In a survey for *The Guardian* in 2016, 82 percent of teachers in the UK said their workload was unmanageable. Almost a third said that they worked more than 60 hours a week.[19] The Education Support

Partnership conducted a survey that revealed that 80 percent of teachers had suffered a mental health problem in the previous two years.

Class sizes are a persistent issue. In the UK they can readily be over 30. In the developing world the figure could be 60 or higher. Yet the educationalist John Dewey suggested more than a hundred years ago that the ideal class size would be 8–12. Something similar is true in universities, where lecturers pontificate in front of huge audiences, sometimes running into hundreds, while direct and interactive contact time is minimal.

So, the education system is currently massively undersupplied with teachers. As robots and AI add to teaching capacity, the effect will probably not be a wholesale culling of teachers but rather a reduction in class sizes and a reduction in the importance of lectures. Indeed, the potential changes are much more radical than this. As with so much else in the new world, the effects of advances in AI will require educational services to be radically reconfigured.

Sir Ken Robinson has a nice turn of phrase to describe the current structure of education: "Students are educated in batches, according to age, as if the most important thing they have in common is their date of manufacture."[20] Surely, the way forward is for tuition to be more personalized and more interactive through seminars and tutorials. Once this has been achieved, the well-entrenched problem of one-size-fits-all education will have been solved.

Ironically, the age-old method of learning at Oxford and Cambridge, that is to say, through one-to-one tutorials (called supervisions in Cambridge), which has been widely decried as old-fashioned and set to go the way of all flesh, could readily be the way of the future, and not just for elite institutions but also much more widely.

AI can provide pupils with individually tailored lessons, effectively putting an end to inefficient group learning. Teachers would be freed up to engage in more active one-to-one lessons. This could even lead to an *increase* in the demand for teachers even as robots and AI take over the traditional pedagogic role of transmitting information.

While technology can be used to engage students and facilitate more personalized learning, it can also offer students who, for a variety of reasons, previously could not receive a conventional education, an online platform to gain access to educational material. As Ed Tech expert Donald Clark has put it: "Google is the greatest pedagogical success and it is a piece of AI."[21] There are now companies that can create customized textbooks. An interaction between teachers inputting the requirements of the syllabus, information about the individual students, and the AI algorithms can produce material especially shaped for the needs of individual students.

In a speech in 2012 Michael Gove, the then Secretary of State for Education in the UK, praised how games and interactive software can captivate students, making learning enjoyable, helping them acquire complicated skills and rigorous knowledge. Furthermore, adaptive software can personalize teaching processes in response to different pupils' understanding.[22] Not only can these interactive systems help students, but they can also help teach subjects such as computer science, where teachers might lack knowledge.

Although we can be sure that the use of AI in education will increase, and we can be fairly sure that the effects will be beneficial, we cannot be at all sure about the magnitude of the impacts. That hasn't stopped some people and institutions from gushing about the possible effects. The organ eSchool News announced in one of its recent studies that the use of AI in the "education industry" will increase by 47.5 percent by 2021.

A lifetime in the forecasting business has taught me that a forecast of anything in three years' time involving a decimal point isn't worth the paper that it is written on – or, more likely today, isn't written on. Anything – the incidence of death by dangerous driving, rainfall in Central America, the number of births out of marriage – we cannot know with any certainty, never mind to the precision of a decimal point. In this instance, even a statement that the use of AI in education will increase by about a half would be heroic enough.

What AI can and cannot do

As well as facilitating a reduction in class sizes, Sir Anthony Seldon sees AI eventually being able to instruct students, mark work, and write reports. Is he right? The marking of simple question-and-answer exercises or multiple-choice questions can easily be done by

AI. Some enthusiasts even suggest that AI will soon be capable of effectively grading essays. Here I have my doubts, except for the most basic of questions and except for marking in the most basic sense. At a higher level I cannot see AI being able to replace human assessors – at least until the Singularity, which is due to hit you in a couple of chapters' time.

During my time at Oxford there was a story (quite possibly apocryphal) of an entrance exam question in philosophy that was, quite simply, "Is this a question?". In the story, one brave student answered simply, and solely: "If it is, this is an answer." How could any sort of AI mark that? Actually, even the dons at Oxford must have found it difficult! If I were responsible for marking that exam, I certainly wouldn't want to fail this student. He or she might be ignorant and lazy, or he/she might be outrageously brilliant (as well as brave). At the very least I would want this student to be interviewed – and not by some form of AI.

Another of my favorite philosophy questions was as follows: "A man is about to make a hazardous journey across the desert, where there will be no water. He equips himself with a gourd of water. While he sleeps, one of his enemies poisons the water. Another slits the gourd so that the water drains out. Who killed him?". Again, I would not want the answer to that question assessed by AI.

Intriguing questions are not restricted to philosophy. In economics, one of my favorite questions that I encountered at Oxford is: "Was the Black Death a good thing?". Answering this question did not require a detailed acquaintance with what happened during the Black Death of the fourteenth century, nor any general knowledge of medieval Europe. All you needed to know is that the Black Death wiped out a huge proportion of Europe's population. The question simply required the student to deploy their understanding of economics to analyze the consequences. I cannot imagine any form of AI being capable of assessing adequately the range of possible answers to this question.

And AI has more general limitations. Teaching has a distinctly human aspect to it. It involves a special sort of empathy between teacher and student. That is why it can bring such rewards and satisfaction for those who have a true vocation for teaching. And on the other side of the relationship, students can thrive on the human interaction. I know that I am far from alone in being able to trace my rise to notoriety, or whatever else I have achieved, to one particularly inspiring teacher. I owe him everything. (What subsequently went wrong I blame on myself.) However good a robot is at transmitting information, it cannot possibly replicate this effect. And for a very good reason – or rather two. Empathy is a human quality. So is inspiration.

The human element is important in all subjects, but it is probably especially important for the nonacademic subjects such as art, music, drama and sport which, as I argued above, are central to a good, rounded education. With the help of AI, teachers will have more time freed up to devote to these nonacademic activities.

How much education?

Most of the above points about subjects and methods apply to all stages of education. But there is a particular question about tertiary education that demands some attention, namely, "Is there too much of it?".

Tertiary (i.e., post-school) education was ripe for fundamental reform even before the AI revolution but the changes to education made possible by robots and AI offer an opportunity to rethink what education is for and how effective the current system is in meeting these objectives. If society fails to consider these questions, then the AI revolution may serve to increase an already serious waste of resources. In the UK, starting with Tony Blair's Labour government, there has been a drive to ensure that 50 percent of kids go to university – a dramatic rise on the 5–6 percent that went to institutions called universities in my day (1970). Admittedly, we then had other institutions called polytechnics, but even including those, the proportion of young people going on to post-school academic education (universities and polytechnics combined) was only about 14 percent. Actually, in the UK by 2016 the proportion of young people under 30 going to university had risen to 49 percent, only a whisker behind Tony Blair's target. In 2017 the education expert Nick Hillman suggested a target of 70 percent university participation rate by 2035.

The idea behind the Blair government's expansion was that, since, on the whole, graduates earn more (and therefore presumably contribute more to the economy) than nongraduates, if you increased the percentage of young people going to university, you would increase the earnings of a good number of people who otherwise might be toward the bottom of the earnings heap, while simultaneously raising national productivity.

This reasoning was always faulty. Just because, on average, graduates tend to earn more than nongraduates, it does not follow that if you create more graduates, they will earn more than they would have done without their degree. Much of the educational system involves coaching students to cross arbitrary hurdles with little or no relevance to productive potential. Accordingly, getting more young people to learn how to jump these hurdles will achieve next to nothing. Still less will be achieved by getting them to jump a rather easier set of hurdles.

This issue goes deeper than the recent overexpansion of tertiary education. Indeed, it goes right to the heart of all education. Does what teachers achieve for their students increase the overall productive capacity (or, if you like, the happiness capacity) of society? Or is what they are doing merely advancing the interests of their Johnny or Jane against the interests of other Johnnies or Janes being taught by other teachers? In my book *The Trouble with Markets* the terms I used to delineate this distinction were "creative" and "distributive."[23]

A good deal of what goes on in the economy, especially in the financial sector, is essentially distributive. That is to say, it affects who wins and who loses, and therefore the distribution of income across society, but does not affect the overall total of income (except by reducing it through the employment of scarce resources in socially useless activities).

Much the same is true of education. Suppose that inculcating in their students a knowledge of dates and kings and queens adds not one jot to an individual's productive capacity. Nor, by extension, to the productive capacity of society as a whole. Nevertheless, by helping "their" student to pass his/her history exam and thereby qualify for entry to a good university, a good teacher may improve his/her chances of getting a good job. Both teacher and student may feel good about this. Yet, by assumption, in this example such activity is purely distributive, rather than creative.

Arguably, a large proportion of current educational activity is essentially distributive in character. It adds very little to the individual student's productive capacity, whether the student succeeds at a particular course of study or doesn't. Nevertheless, success serves to separate the "winners" from the "losers." Much of the educational process consists of a series of arbitrary hurdles which, if successfully traversed, prepare the student for the next set, and so on and so forth until the "finished article" emerges at the end, specialized in traversing these various arbitrary hurdles that bear little relationship to what the hurdlers will encounter in the rest of their lives.

The case for less education

Given the preponderance of "distributive" activities in the educational process, you could say that in contrast to the prevailing view that there needs to be more education, in fact there needs to be less. A recent book by Bryan Caplan, a Professor of economics at George Mason University in the USA, entitled *The Case against Education*,[24] argues precisely this. He presents two pieces of evidence that the modern education system is essentially about "credentialism" rather than the creation of human capital. First, he says, students who drop out of university just before taking their degrees earn much less than a newly minted college graduate, indeed hardly much more than a high school graduate. Yet, if it was the innate worth of what they had learned that really mattered, the difference should be minimal.

Second, many students do the bare minimum of work to secure their "credential," aware that there is little if any value in the knowledge or skills that the educational system is trying to impart to them. He says that 50 years ago the typical American student spent some 40 hours a week in class or studying. That is now down to 27 hours a week.[25]

Caplan's critique of the educational system is too damning. Even if the successful cramming of knowledge that will soon be forgotten, and was anyway mostly useless, does not add anything to an individual's ability to do a particular job, this does not mean that it is totally useless. The skills associated with learning, and even cramming, and then disgorging the information in a predetermined way at a specified time, can be useful in many aspects of a working career. And doing this successfully requires discipline and determination. Moreover, employers do need some sort of credentials in order to help them make their selections.

But Caplan is surely right that credentialism has gone much too far. The consequences of the Blairite policy of educational expansion have been quite different from what was intended. Just giving a young person a degree does not make them more capable, or more valuable, in the workforce. And the whole experience costs a bomb. So now the labor market is awash with kids in their early twenties with meaningless "qualifications" who either cannot find a job, or can only find one that is not of "graduate quality." The result is that for many "nongraduate" jobs, having a degree has now become a key requirement, even though it adds nothing to the candidate's ability to do the job.

A recent report from the Chartered Institute of Personnel and Development said that 58 percent of UK graduates were in nongraduate jobs. This compares to a figure of only 10 percent in Germany. One unfortunate consequence is that many people who do not have degrees but who could do these jobs perfectly well are excluded. These young people are disproportionately from less advantaged homes. So, this "degree inflation" is far from being an innocent game. Not only does it waste huge amounts of money, but it also helps to accentuate inequality and to propel it down the generations.

In the UK, for the privilege of acquiring a degree that is near-useless in its intrinsic worth and which just enables its holder to squeeze into what used to be regarded as a nongraduate job by squeezing out current nongraduates, young people leave university weighed down with debt. And this problem is not restricted to the UK. In the USA, in 2018 total student debt exceeded $1.5 trillion. According to a recent Brookings study, by 2023 almost 40 percent of these borrowers are likely to have defaulted.[26]

This whole system is an expensive disgrace. We should be prepared for far fewer young people to go to university. But this does not mean that young people should not learn or continue to be educated after school. They may well need to do more "on the job" learning, combined with short courses teaching them particular skills. And they may well benefit (although not necessarily financially) by taking short breaks for academic study during various stages of their working lives.

Artificial Education

The AI revolution can help to facilitate this outcome, but in order to do this it must not simply fit into the existing educational system. In the robot- and AI-driven world it is essential that new and increased educational capacities are directed toward increasing the creative, rather than the distributive, aspects of education. There will be no point in replacing one (socially useless) set of artificial hurdles with another more technologically sophisticated set, designed simply to sift out those better able to jump these hurdles from those less able to. The robot and AI revolution needs to be the catalyst for a complete rethink of education, what it is for, and how best to achieve agreed educational objectives.

The creative versus distributive distinction has a surprising implication for the balance of education between academic and career enhancing studies on the one hand and studies that are designed to enhance leisure on the other. Whereas a good deal of the effort put into academic and career-focused education is partly and, in some cases, largely distributive in character, teaching designed to enhance students' capacity to get the most out of their leisure time is essentially creative in nature. After all, if exposure to Balzac and Beethoven helps a student to escape from *Celebrity Big Brother* and *Keeping up with the Kardashians*, that is a benefit secured for them without a corresponding loss to anybody else's capacity for enjoyment.

Interestingly, out of all developed countries Switzerland sends one of the lowest percentages of its youth to university. You wouldn't want to call Switzerland unsuccessful. Its workforce has a very high skill base, and it enjoys one of the highest levels of average real income in the world. This is surely related to the fact that for those who do not go to university there are institutions that provide high-grade vocational and technological training.

Incidentally, in case you think that this approach betokens a damaging downgrading of academia, it is noteworthy that Switzerland is host to the only European university outside the UK that regularly ranks among the world's finest – Zurich. By comparison, the formerly great universities of France, Germany and Italy are nowhere. We have a lot to learn from the Swiss.

Where and when

As part of the AI revolution we will need to think radically about the conventions of education, such as the length of the school day, the length of the academic year, and the length of school holidays. In addition, we will need to reconsider the length of degree courses which are normally three, or at most four, years, with only about a half of this time actually spent at university. Why? Surely there is scope for more intense degree courses of one to two years. After all, because AI will enable learning to be more personalized, it can be much quicker.

Nor is it obvious that education needs to take place in physical entities called schools and universities. Since so much education will take the form of students interacting with some form of AI-driven educational program, this could just as easily be done at home. In any case, home-schooling seems to be on the rise. In 2016/17 up to 30,000 children in the UK were home-schooled by parental choice. In the US the equivalent figure is about 2 million.[27]

As for university education, the UK's Open University long ago pioneered remote learning, well before advances in AI made this easier. Moreover, many universities have developed degree courses that can be undertaken partly, or wholly, remotely.

Nevertheless, there should be a limit. I have argued above that a full and proper education involves the development of human potential and human personality. This must involve social interaction, as well as sport, drama, and other activities. This requires some shared physical space that we might well want to call a school or university.

But in the Robot Age there will be no need for all, or even the majority, of the time devoted to education to be spent in such places. Rather, the bulk of educational time can be spent at home or in the work place, interspersed with occasional visits to the physical "seat of learning." Making sense of this will involve a fundamental rethink of how we use educational resources – including buildings and physical infrastructure.

Constant learning

The title of this chapter, "How the young should be educated," is in some ways inappropriate because it is not simply the young who need educating. The traditional model for education and training is that this occupies the first 10–20 years of life after infancy. A period of about 40–50 years of full-time employment, and not much further formal education, or learning or training, follows on afterward.

This model has a certain logic to it when the economy is static or when, even though economic output is growing, economic structures and the demand for certain skills are reasonably stable. In fact, we have not been in such a world for quite a while yet the relationship between work and education has trundled on as though nothing had changed. The AI economy promises to be much less like what we have grown up with.

The implication is that people will need to have several periods of learning and training in their life, interspersed with periods of employment.[28] The great thinker and futurist Alvin Toffler long ago suggested that "the illiterate of the twenty-first century will not be those who cannot read and write, but those who cannot learn, unlearn, and relearn."[29] This implies that the labor force must be able to change as rapidly as the environment around them is changing.

This can apply as much to "education for leisure." There is no good reason why our investment in our ability to enjoy ourselves and derive deep pleasure from our leisure time should be restricted only to the first 10–20 years of our life.

Mind you, the advances of technology are not necessarily all positive for the education and overall mental wellbeing of citizens. There is some evidence of cognitive decay as a result of recent technological developments, notably in video game technology and smartphones.

In his book *The Shallows*, Nicholas Carr argues that the internet is having an adverse effect on our ability to think.[30] In an article published in *The Atlantic* in 2013 called "All Can Be Lost: The Risk of Putting Our Knowledge in the Hands of Machines," he bemoaned the rise of "technology centered automation that elevates the capabilities of technology above the interests of people." Similarly, it has been argued that reliance on GPS for directions is hindering our ability to reason and remember spatially.[31]

If it were to be established that, above and beyond some threshold level, exposure to these things did indeed impair cognitive ability, then there would be an argument for restricting humans' reliance on such technologies. Indeed, perhaps a key role for the educational process would be to teach people how to get the most out of the new technology without becoming addicted to it, and without suffering cognitive decay.

The role of the State

What should state policy be with regard to education and training in the Robot Age? I have already discussed the implications of the spread of robots and the development of AI for the type of education that will be most profitable and desirable, and which educational methods will work best in the new conditions. In principle, shifting education in this direction can be accomplished by normal market forces as parents and their children, as well as teachers and educational experts, come to appreciate what the new world will require and alter their educational preferences accordingly.

Yet in virtually no society in the modern world is education left entirely to the private sector with no role for the state. There are two good reasons for this. The first is that, as we discussed earlier, education probably has strong positive externalities, that is to say, not all the benefits of someone being educated fall to that individual. Some are spread more widely across society. If you like, society has a self-interest in its members being educated. In other words, if education were left purely to private provision, not enough of it would be provided for the public interest.

This is partly due to imperfections in the markets for the provision of finance for education but also partly due to myopia and lack of information about the prospective benefits of education among the poorer (and usually least well-educated) parts of society.

Admittedly, there are some thinkers, such as Professor Bryan Caplan, referred to earlier, who argue that education may give rise to *negative* externalities. To the extent that they are right, there is a case for *reducing* state funding of education. Indeed, Caplan argues precisely that. He says that he favors the full separation of school and state. But, as I said above, Caplan exaggerates his case.

Moreover, even if Caplan were substantially right, his radical prescription comes up against the second argument for state involvement in education – the distributional element. Without some form of public provision or financial involvement, education would tend to become the preserve of the better-off in society and this would accentuate division and inequality. Indeed, such inequality would cascade down the generations.

I am going to discuss policies to reduce inequality in the next chapter. But perhaps the best thing that the state can do to reduce inequality in the AI economy is to invest in education. Actually, this was true before the onset of the AI revolution, but it is even more true now.

There are several ways that an effective education policy can contribute to the lessening of inequality. Most importantly, it can counter the advantage that the children of well-off and successful parents have in being sent to private schools, enjoying private tutoring or simply absorbing educational benefits in the home. A radical reform of the state educational system, complete with much more public money, would seek to raise educational standards in publicly funded schools to those prevailing in the private sector, including in nonacademic subjects that will help to equip people properly to be able to use their increased leisure time fully and "profitably."

By appropriate structuring of what is taught, the state can equip young people with the skills and attributes that will help them to find and keep rewarding work in the labor markets of the future. But the state can also contribute to the funding of life-time learning and retraining, thereby again helping people to reequip themselves with useful skills. This is particularly important as a way of ameliorating the conditions of all those millions of people who will, from time to time, suffer from the gales of creative destruction blowing through the economy, including as a result of the spread of robots and AI.

Most importantly, in order to reduce inequality in the AI economy, the education system needs to help produce the kind of workers who will thrive as complements to AI. The MIT professor David Autor makes the point that in 1900 the typical young, native-born American had only a very basic education. As agriculture was in sharp decline and industrial employment was increasing, children would need additional skills. The US government responded by being the first nation in the world to offer universal high school education to its citizens. What is really noteworthy is that the movement behind the offer of universal high school education was led by the "farm states."[32]

Of course, individuals themselves will have a self-interest in undertaking such lifetime learning and retraining. But they may not be in a strong position to fund it adequately. And there are externalities involved in ensuring that people are employable. Nor can we rely on employers to invest in lifetime learning and retraining – at least not to the necessary extent – because the possibility of employees moving to another firm, or of the employing firms going bust before a full payoff of the investment, mean that they will not be able to capture the full social benefit of such investment. Accordingly, they will underprovide.

Mind you, making progress on this agenda is easier said than done. It is no use just pouring money into the existing setup. As I have argued above, the structure of education needs to be fundamentally reformed – everything from primary and secondary education, through university education to lifetime learning and retraining. Such radical reform has already encountered strong opposition from the teaching profession. This will continue and even intensify. The educational system is one of the most conservative parts of modern society. Opposition to change from vested interests in education is another reason why the state needs to be actively involved in nudging the system into adapting to new requirements.

Nevertheless, a word of warning is in order. Complete reliance on the state is certainly not the answer. In a powerful sense, each person is his or her own educator. AI greatly increases the scope for self-education. And beyond individuals themselves, there are other family members from whom we can learn so much. Governments around the world frequently make the mistake of equating education solely with what goes on in schools and universities or in centers of workplace learning.

Moreover, it has been noteworthy that state-funded education has tended to overemphasize learning and the achievement of certain paper qualifications over "fuzzier" noncore activities such as music, drama, and sport. Yet, as explained above, the qualities that are developed through such activities are increasingly the ones that employers are looking out for. They tend to find them, disproportionately, in the products of the private educational system.

So, it isn't only the teachers and educators who have to change, but also all those involved in planning and implementing state-funded education. In the AI economy, as well as providing a good academic training, the ambition of state-funded education should be to provide students with as good an experience of nonacademic activities as the private sector. This can both further their development as human beings and promote their employment prospects.

Conclusion

Education is yet another example of a phenomenon we have encountered throughout this book. AI enthusiasts gush about the way robots and AI are going to transform a particular aspect of society – in this case, education – but they completely misjudge the nature of that transformation. Supposedly, teachers are bound to go the way of taxi and lorry drivers, set to be made redundant and consigned to become yet another lump of unemployed skilled workers having to scrabble around for some new way of earning a living. Meanwhile, the subjects that need to be taught are revolutionized, with traditional arts subjects thrown out of the window, to be replaced by STEM subjects only.

The reality is likely to be very different. Education was badly in need of reform before the advent of robots and AI. Its practitioners and protectors are some of the most conservative people on God's earth. Quite apart from the changes to be unleashed by robots and AI, they have been long overdue for a shock.

The AI revolution will deliver it. But it will bring changes very different from the ones that the AI radicals suggest. Liberal arts education will not die but will remain as important, if not more so. The aim of education will go beyond the fashioning of people for the labor market and will involve helping people to become rounded individuals, able to make the most of their leisure time, as well as improving their careers and their contribution to society.

And the number of people employed in education is not set to plummet, as the radicals suggest. Indeed, it may well increase. Rather, robots and AI are set to transform educational methods, with the result that education is more personalized, and teaching is no longer delivered though the mass-production approach, reminiscent of a car factory.

Much of this agenda for radical change will be delivered naturally through the market, with schools responding to market pressures and the preferences of parents. Yet both teachers and parents will doubtless take time to adapt and to become confident that the robot and AI revolution does not imply the end of liberal arts education or a lessening in the importance of nonacademic subjects. This is an area where the market cannot be relied upon to deliver the desired result unaided. After all, a large part of the educational system is both funded and controlled by the state. So, it is vital that both the educational establishment and political leaders understand the true implications of robots and AI for education, and are not duped by the AI hype.

Getting the education system right will, if anything, become more important for the health of society because it will be critical to the outcome of another of the main challenges facing us in the Robot Age, namely, how to ensure that large numbers of people are not left behind.

9
Ensuring prosperity for all

> "We will eventually need to separate money from occupation, and that shift will challenge some very central assumptions about how we define our own values and identities."
>
> *Gerd Leonhard*[1]

> "Something has to be done. This is something. So it has to be done."
>
> *Anonymous*[2]

In Chapters 3–6 I argued that there is no reason to believe that the Robot Age implies the Death of Work. There will be plenty of jobs for people to do – although increasing wealth will probably see a fall in the number of hours that people, on average, devote to work rather than leisure. And in Chapter 5, I laid out my thoughts about what sort of new jobs may come along to take the place of the old.

Even so, it is possible that the jobs that are available for many people are temporary, insecure, and badly paid. Perhaps the employment market in some poor countries today provides a vision of the future for all of us. When you arrive at the airport swarms of people offer you services of dubious value, such as carrying your suitcases when a baggage cart would do the job just as well. And at hotels, and even sometimes at restaurants, armies of flunkies are there to do your bidding, but with the creation of hardly any value for you and consequently the earning of hardly any money for them.

Alternatively, or perhaps additionally, as I suggested in Chapter 5, the future employment market may resemble premodern times in the West when large numbers of people worked in domestic service and the really rich employed armies of retainers (loyal or otherwise).

These are two visions of the future but, as I argued in Chapter 6, they are not the only ones. Indeed, I laid out there several factors that may retard or offset any tendency toward increased inequality. At the very least, this means that we should not rush headlong into adopting a radical program for income or wealth redistribution that may be both unnecessary and damaging.

Nevertheless, I could be mistaken about how robots and AI will affect income distribution. If I am seriously wrong, then we could end up with a distribution of income that many of us would find unjust and unacceptable, with unpalatable consequences for the shape and nature of society. This is indeed the vision of the future that haunts many AI experts and others. So we need to think about what to do if this nightmare becomes reality.

What could public policy do to prevent it and/or to tweak the distribution in a more egalitarian direction? Should we, and can we, rely on the existing systems for income redistribution, perhaps suitably reformed and improved? Are there radical new measures that could be adopted? And, if there are, should we embrace them? In what follows, I will give special attention to the idea of introducing a universal basic income (UBI).

But before we consider ripping up the current system of tax and redistribution, we need to go back to first principles. Putting robots and AI aside for a moment, what sort of distribution of income is "good"? Or at least acceptable? And how far does the current system for redistributing income achieve these goals? It is only when we compare any putative new system with the existing one that potential policy measures for the AI economy will have cogency. It is vital that we have a clear understanding of these issues because society is a complex organism and major intervention to alter its mores, customs, and institutions can have serious unintended consequences.

The ideal distribution

Philosophers and political economists have pondered and waxed lyrical on the subject of the ideal distribution of income since ancient

Greece. It would be unkind to say that we are still none the wiser. But it would be true to say that there is no settled conclusion on the issue.

Marx produced a striking vision of an egalitarian future under communism, which would bloom only after an agonizing period of inequality and injustice under capitalism. His biting analysis and stark conclusions derived from the concepts of "surplus value" and "exploitation." Unfortunately, although these concepts seem strong, in practical terms they are brittle and break in your hands. As for Marxism, in practice, I will not berate you here with another account of its horrors.[3]

Perhaps it is unwise, and even impossible, to identify the fairest distribution of income. We should simply stick to identifying egregious examples of *unfair* distributions – and doing something about them. When we see great wealth and comfort sitting side by side with acute poverty and deprivation in the streets of Mumbai or the favelas of Latin America, we know that something is wrong. But less extreme contrasts grate, too: the superyachts of the superrich bobbing along idly, while millions of ordinary people slave away at exhausting jobs just to keep their modest little homes afloat.

These sharp contrasts may make us uncomfortable even when the superrich have "earned" their squillions, but we encounter further challenges when the superrich derive their riches through inheritance while the poor derive their poverty from their un-inheritance.

These comparisons of income and wealth are not simply about issues of justice and morality, or even greed and envy. They are also about a certain sort of efficiency. Economists have the concept of the "diminishing marginal utility" of income. Simply put, this means that as you get richer, each extra bit of income brings you less and less benefit. This surely accords with our own experience. The first bit of extra income that enables you to buy enough food to fend off hunger is worth far more to you than the equivalent bit of extra income when you are richer, which may enable you to eat out rather than preparing the food yourself.

From this realization, it is not a great jump to argue that an extra pound/dollar given to a superrich person is worth far less than an extra pound/dollar given to someone on the breadline. And, from this, it is no jump at all to argue that a pound/dollar taken from a superrich person brings a far smaller loss to them than the gain experienced by this pound/dollar being given to the poor person.

Following this logic, if you were trying to achieve the objective laid down by the famous philosopher Jeremy Bentham of pursuing "the greatest happiness of the greatest number," you would aim for complete equality of income, or something very close to it.

But, of course, this is a very crude view and it leads to very crude conclusions. In practice, we can never know how much a dollop of money or expenditure is worth to one person versus another. Accordingly, the objective of just adding up each person's "utility" to reach a grand total for society as a whole rests on false foundations. Moreover, society is more than a collection of atomized individuals (or even families) maximizing their "utility." Anyone trying to promote human happiness must recognize society's complexities and the complicated motivations and interactions between individuals and institutions.

Conflicting principles

In practice, there are several opposing principles that must be taken into account when even a philosopher-king, never mind a modern politician, seeks to equalize the distribution of income. People are very different in their talents, inclinations, and efforts. Complete equality of outcomes would therefore be unfair. It would also be resented by most of those people who would otherwise be toward the top of the income pile.

Moreover, people respond to incentives. The ability to improve one's lot is an incentive to effort. Accordingly, a society of enforced equal outcomes for all would be poorer than a society where people could enrich themselves. And people are also to some extent motivated by relative position. Those who are ahead are motivated to stay ahead and those who are behind are motivated to catch up. So again, an equal society would be a poorer one.

In addition, people have a deep-seated instinct to pass on what they have earned to their children. Stopping them from doing so would lead to unhappiness. It may also undermine motivation.

More fundamentally, sometimes different life outcomes are the result of good or bad fortune, such as winning the lottery or suffering from a car accident. While the good society might want to prevent people from suffering from the latter, it could be a diminishment of human life to rule out or restrict the former.

Indeed, if an attempt is made to "correct" the distribution of income, someone has to be responsible for it. That "someone" is the state. The redistribution process is then subject to all the vagaries of the political process, with the result that the outcome falls well short of what a philosopher-king would devise.

Moreover, this process may strengthen the power of the state not only against the individual but also against the other institutions of civil society. This may not matter if the state is in the hands of philosopher–kings but it most certainly does when it is in the hands of flesh and blood political leaders. Accordingly, the objectives of ensuring the preservation of freedom and resisting over-powerful rulers are probably best served by having some people in society who are independently wealthy and can therefore resist the blandishments of state power. (Mind you, this offers no guarantees. There are several states in the world today that amount to tyrannies, despite the presence in their societies of umpteen billionaires.)

This set of principles underlies a reluctance in modern society to enforce a fully egalitarian outcome. They must be borne in mind when we consider the impacts of the AI revolution on the distribution of income and any policy measures that may be undertaken to change that distribution. Even so, there are two key longstanding arguments that point in the opposite direction, and they gain strength from the AI revolution. They deserve some attention now. In addition, there are two other arguments that have come to prominence more recently, particularly in connection with robots and AI, that I will come to in a moment.

First, and most fundamentally, it can be argued that someone's abilities aren't really "theirs" at all. Are differences in talent, inclinations, and effort really down to the individuals themselves? Aren't they really also, at least partly, the result of inheritance? In the current world, never mind in the coming world where beauty may be even more prized than it is now, when a beautiful person becomes a megastar in reality TV, causing them to make squillions, is their "talent" really theirs, as opposed to being the product of their inheritance, and therefore subject to all the usual objections to inherited wealth? Isn't this also true of those people blessed with brains, or even a strong work ethic? Accordingly, you can readily argue that the recipients of large incomes based on "talent" do not really "deserve" them.

Second, it is striking that people performing the same jobs in countries at very different levels of development earn very different incomes. A bus driver in Entebbe, Uganda will earn much less than a bus driver in Berlin, even though they are doing the same job. The same is true of doctors and, indeed, just about all categories of job. In other words, what an individual earns is not just the product of their own efforts and skills but is also substantially the result of the society in which they live. Accordingly, for those who live in rich, successful countries, a good part of these rewards properly "belongs" to society more generally.

The modern compromise

So, how to reconcile these conflicting principles? The only answer is some form of compromise. Or, to put the matter less kindly, a fudge. This has been the response of societies since time immemorial, well before robots and AI were even a pipedream.

There have been occasional and ad hoc measures to enhance the position of the poor throughout history, but nothing compared to the scale and ubiquity of redistributive polices undertaken by modern states. The change can be said to have started with the measures enacted by Chancellor Bismarck in imperial Germany in the late nineteenth century, namely the institution of state-funded old age pensions and unemployment and sickness benefits.

The role of the state came to full fruition in the years after the Second World War in the development of the welfare state. This is most advanced in the countries of Europe, but there are elements of it operating in North America, Asia, and all other parts of the world. Essentially, the welfare state aims to be a compromise solution to the competing arguments for total laissez-faire on income distribution and the pursuit of an egalitarian outcome, discussed above. It has several components:

- *The provision of certain public services either free or at a highly subsidized rate.* Since spending on these services takes up a higher share of expenditure for poorer people, their provision at lower than market prices serves to tilt the income distribution in favor of poorer people. These services typically include education, health, and, in some cases, public transport. Typically, they do not include three other key categories of expenditure that are heavily weighted in the share of expenditure by poorer people, namely food, clothing, and heating (as well as other utilities). Since state-provided or -subsidized services are funded out of general taxation, provided that the tax system disproportionately bears on the better-off in society (see below), this buttresses the redistributive effect on post-tax disposable incomes.
- *The provision of subsidized housing for those with low incomes.*
- *The provision of unemployment and sickness insurance.* As originally envisaged, these benefits were indeed thought of as a form of insurance, with the state collecting premiums through a form of taxation and then disbursing benefits as the need arose. In practice, however, the insurance element has been watered down as governments have not kept separate funds and benefits have to some extent been available whether or not contributions have been made.
- *Ad hoc benefits for people falling below certain specified levels of income.* Typically, these benefits have been paid when a family's income falls below a specified level and are therefore given up as and when income rises above that level.
- *State-funded top-ups for income earned through low-paying employment.*
- *The payment of some benefits if certain criteria are met without subjecting the beneficiaries to means testing.* Examples include old age pensions, free pensioner travel, zero-cost television

licenses for people over 75, allowances to pensioners to cover the increased cost of heating during the winter months, child benefit, and disability benefit.

- *A tax system that includes a tax-free allowance and/or a lower rate of tax at low incomes.* In most countries, the tax rate, both marginal and average, also rises as income increases. (Mind you, in recent years there has been a move toward flatter taxes. Hong Kong and Russia, among other countries, operate a system in which the tax rate never rises above the standard rate, no matter how large the income.)
- *The taxation of investment income, even though the income from which money was saved in order to accumulate capital was itself taxed.*
- *The taxation of inherited money even though it has already been taxed when it was earned.*
- *In some countries, the taxation of wealth.*
- *Encouragement of charitable giving through tax breaks for registered charities.*

This system is chaotic. Even without any of the pressures that are about to be unleashed by the AI revolution, it is surely ripe for reform. It has several major weaknesses. For a start, it is complex, difficult for people to understand, and expensive to administer. Moreover, in most countries individuals do not have a sense of ownership of their accumulated state-provided pension rights and other entitlements to benefits and cannot access funds flexibly. (Singapore is an exception.)

Much of what goes on involves the state taking with one hand and then giving back to the same people with the other. Since the administration of this money-go-round is expensive, and since marginal tax rates have to be higher to fund the expenditure, this is massively wasteful.

It is doubtful whether the state should be in the business of providing insurance. Even if some guarantees and top-up funding are required, arguably the insurance element is better provided by the private sector. (Admittedly, though, with regard to health insurance there is a problem of adverse selection. That is to say, the health insurance companies are reluctant to insure those most likely to need medical care. This requires state intervention in some form or other.)

Moreover, state-provided services, in health and education, have a patchy record. Critics would argue that they have a tendency toward poor outcomes in regard to quality, consumer choice, and efficiency.

The benefits system embodies strong disincentives to work as benefits are withdrawn as income increases. In some cases, people face a marginal tax (and benefit withdrawal) rate of 100 percent or more. Meanwhile, the tax system is riddled with loopholes. Often while it manages to extract a considerable proportion of middle incomes in tax, large fortunes escape much more lightly as their owners benefit from international mobility and are able to afford the best tax advice.

Indeed, in many countries the system is so riddled with loopholes and anomalies that the rich and superrich pay lower percentages of their income in tax than the legislators intended, and sometimes "less than their cleaners." Meanwhile, major international companies manage to pay very little corporation tax, while small domestically rooted companies have no option but to pay the full whack.

Perhaps most importantly, the overall cost of the welfare state is huge and set to get even larger as populations age. Accordingly, the disincentive effects from the high marginal tax rates apparently needed to fund the system are set to get even greater.

This system has grown up in a *higgledy-piggledy* way thanks to the interplay of the desire to preserve the contributory principle, and to help people only when they need it, while giving benefits to certain "deserving groups" (e.g., parents, pensioners or the long-term sick), while wanting to limit the cost and any adverse effects on incentives. It badly needs to change.

Enter robots and AI

But what difference does the AI revolution potentially make to this issue? The first point is surely that however uncomfortable we may be with the current income distribution, and even more so with the distribution that would obtain in the absence of redistributive polices, if the pessimists are right, it may well become more unequal as AI and robots exert a greater influence in the economy. Moreover, there is a chance that, as well as skewing the income distribution, the

AI revolution may reduce social mobility and hence have a tendency to propel the unequal income distribution down the generations.

Furthermore, such an unequal distribution has two significant unwelcome side effects. It may tend to depress aggregate demand as spending power is concentrated in the hands of those least inclined to spend it (as analyzed in Chapter 3) and thereby put added pressure on demand management policies whose effectiveness is less than perfect. Moreover, it may tend to undermine democracy (as analyzed in Chapter 7).

If any of these outcomes transpires, and if we wish to resist this result, then we need either to ratchet up the current set of policies designed to redistribute income or develop some new ones.

Mending the current system

Suppose we took the decision to mend the current system for redistributing income. What could we do? In what follows here I give an outline of what, in principle, could be done. This provides some context and counterweight to a much more radical measure, namely the introduction of a universal basic income (UBI), that I will consider in a moment. But I do not wish to imply that the measures I discuss here for reforming the existing system are either politically easy or necessarily desirable. Nevertheless, they are feasible – at least in principle.

First and foremost, the state could undoubtedly become more efficient in its policies for redistributing income. It could improve the efficiency of public services and hence deliver better results for the same money, which would disproportionately benefit the less well-off. Education is particularly important here because it has such a bearing on the earning capacity and life chances of people who are born to parents without much money. Hence it can have a major bearing on social mobility. (In Chapter 8 I discussed some of the ways that the state could use education policy to influence the distribution of income.)

In many countries across the world which operate with a low share of government spending and taxation in GDP, including the USA, Japan, Switzerland, and many others, it would be possible to increase the ratio of spending and tax to the levels currently operating in the higher spending European countries, including France and

Scandinavia, with the extra spending devoted to increased benefits for the poor and financed disproportionately by taxes on the better-off.

Several authors have argued that, even without the AI revolution, there is a strong case for a more generous system of social benefits. In 2011, the Dutch Ministry of Health commissioned a study of the costs against the benefits of relief for the homeless (including free shelter, assistance programes, free heroin, and prevention services). It concluded that investing in a street sleeper offers the highest return on investment around. Every euro invested in fighting and preventing homelessness in the Netherlands enjoys double or triple returns in savings on social services, police, and court costs.[4]

Another approach to the reduction of inequality would be for the state to discontinue certain universal benefits, such as childcare and pensions for people above some threshold, and to allocate the money to "deserving cases." It could also increase the taxation of the superrich. Of course, it is open to all countries to increase standard tax rates. In practice, that may not be a sensible way of proceeding. A system of lower rates, with fewer exemptions and no loopholes, may bring in more revenue. Action to tackle tax avoidance by both wealthy individuals and corporations would be feasible but to be effective it would need to be internationally coordinated.

It is also open to countries to impose wealth taxes, or where these already exist, to increase them. Again, though, to prevent such a move being undermined by capital flight, it would have to be internationally coordinated. As things stand, there seems scant chance of getting international agreement on the imposition of a wealth tax.

Competition and reform

A completely different approach to inequality would be to tackle it at its roots – or at least some of them. In the USA, anyway, recent increases in inequality have not been caused primarily by increased returns to capital (as argued by Thomas Piketty and discussed in Chapter 6) but rather by increased disparity between high- and low-earning employees. What's more, a substantial amount of this increased inequality is accounted for by large increases in the pay of

CEOs and other senior corporate executives, and major increases in the relative pay of people working in financial services.[5]

These increases in inequality have nothing directly to do with AI. Nevertheless, if society wanted to do something to reduce inequality then, at least in the USA and the UK, the place to start would be with strengthening corporate governance procedures to bring about lower pay at the top of corporations, plus reducing the size of the financial sector in the economy. In addition, government could introduce a program of anti-monopoly measures, especially in the digital sector, where several companies make enormous profits from their quasi-monopoly positions. This last suggestion mirrors what happened in America under the "Anti-Trust" program in the early twentieth century.

But, of course, none of the above suggestions would be easy, not least politically. And at first blush, to many observers of the AI scene, none of them seems to match the scale and significance of what is about to occur because of robots and AI. Accordingly, there is a groundswell of support for a truly radical approach to income redistribution, which seems comparatively easy, appropriate to the problem at hand and politically feasible, namely the introduction of some form of basic or universal income. The idea has resonance without the possible effects of robots and AI. But, as the ensuing detailed discussion should make clear, it seems to have particular relevance to a world undergoing the robot and AI shock.

A universal basic income (UBI)

The idea of a guaranteed minimum income, or GMI, often referred to as a universal basic income (UBI), which is the nomenclature that I will use here, comes in many variants.[6] In its purest form a UBI is the grant of a regular income at a single fixed level per individual (or per household), regardless of circumstances, financial or otherwise, and without the need to fulfill any conditions, save being a citizen of the country in question, or having been a resident there for so many years.

This last qualification raises a tricky issue. If recent migrants are excluded from the UBI then there has to be some sort of ancillary welfare system to cope with their needs, as and when they fall on hard times. And if they are *not* excluded and are thereby able to draw the UBI then there could be considerable resentment by indigenous taxpayers. Moreover, if one country granted immigrants a UBI and others didn't, or if it granted a particularly generous UBI in comparison to its neighbors, then it could attract large numbers of migrants.

In some variants the UBI is paid from birth; in others, from the attainment of adulthood. In some variants it can rise with age and be set at different levels depending upon personal circumstances – for example the number of children or even the part of the country where the person lives. In some impure variants, it is dependent upon the recipient falling below some income level, and in certain others it requires recipients to be looking for work.

In some variants, the amount of the benefit is indexed to prices or wages. In some cases it even rises automatically with GDP per capita. In still others, the payment of the UBI is limited to a certain number of years per person. It is usually suggested that the UBI should not be mortgageable or taxable, although if it were taxable, provided that the tax system is progressive, this would be a way of clawing back the benefit from better off people.

In the purest version of the idea, the UBI is envisaged as replacing all other forms of welfare; in others, though, it is envisaged as a supplement rather than a replacement for any or all of them. In one variant, the benefit system is merged with the income tax system. In this variant, below a certain level of income, the tax rate turns negative and people receive money from the tax authorities. (This is known as a negative income tax.)

There has even been a proposal that the state should provide a "universal minimum inheritance." It has been suggested that this should be set at the sum of £10,000 for all 25-year-olds.[7] This is a variant of the UBI idea and shares both its advantages and its drawbacks. The proposal was not made specifically with the AI revolution in mind, and it stands or falls without it. The suggestion derives from increasing concern about rising inequality in our society. But the specific concern about the distributional consequences of the AI revolution and what could be done about it mean that the suggestion of a state-funded lump sum seems particularly relevant for the AI economy.

In practice, though, this proposal is all wheeze and subterfuge. In order to fund the handout, there have been suggestions for a levy on companies, new taxes, and the proceeds of state asset sales going into the fund. But these specific features add nothing to the central idea of giving citizens a non-means-tested benefit, financed out of the public purse. There is not a lot to be said for the various ways suggested to raise money for a "fund" to finance the giveaway. We might as well be honest and clear about what is going on and simply finance the benefit out of general taxation.

Nor, if the lump sum and UBI suggestions are alternatives, is it obvious that a lump-sum payment is an improvement on a regular income paid every week, month or quarter. One possible advantage of a lump sum is that it is more compatible with the accumulation and preservation of a capital amount. Admittedly, the recipients could equally amass such a sum by saving up all their regular income receipts. But most people find it difficult to exercise sufficient self-restraint. Yet, equally, some recipients will tend to blow a lump sum in one go, leaving them without the continuing support that a regular income provides.

In what follows, I will leave the lump sum idea to one side and instead analyze payment of a regular income.

Basic jobs

But before we get to the main issue, there are two other variants to consider, namely the idea of a basic jobs guarantee (BJG), and the provision of "universal basic services" (UBS). In the USA Senators Bernie Sanders, Elizabeth Warren, Cory Booker, and Kirsten Gillibrand have pressed for a BJG to be trialed.

This idea aims to head off one of the criticisms of UBI but immediately undermines one of its key appeals. Clearly, the proposal responds to the critique that a UBI would reduce the supply of labor and undermine the work ethic (which I will discuss in a moment). But in order to meet this objection it reasserts the link between work and income. It therefore lacks the appeal that UBI has of allowing some people, including would-be poets, writers, painters, and composers, to have a livable income without having to do paid work.

So far, at least, the idea of a BJG has failed to enthuse key figures generally thought to be on the Left, who might be thought to be sympathetic to it, on the grounds of both cost and doubts as to what people receiving the jobs guarantee would actually do. Lawrence Summers, a former US Secretary of the Treasury, has posited the case where a $15-per-hour guaranteed job encouraged an extra 4 million people into the workforce and attracted 10 million existing employees (for a quarter of whom $15 an hour would represent a wage increase.) He reckons that the cost would be $60,000 per worker, giving a total increase in annual government spending of $840 billion, or about 20 percent of the current total.[8] This would surely be unaffordable and grossly inefficient. Accordingly, we should assume that this BJG suggestion is not going to fly.

Universal basic services

Another idea is to provide a package of universal basic services, including free housing, food, transportation, and communications. This idea has attracted a number of politicians on the Left, including the UK Labour Party's John McDonnell, the Shadow Chancellor of the Exchequer.

Proponents of this UBS idea argue that it is merely an extension of the system already in place that provides free health and education services. They are right about this but therein lies an implicit danger because state provision of these services in the UK has met with widespread dissatisfaction. Moreover, extending the free provision of other things such as food, housing, and transportation would lead to extraordinary waste and inefficiency. It would also, of course, extend the state's tentacles still wider into society. Moreover, it would detract from two of the key attractions of the idea of a UBI, namely its simplicity and the way it leaves the individual free to choose what to do with the money provided by the state.

So, the UBS idea is likely to be attractive only to a relatively small number of people on the Left and will probably not gain wider traction. Accordingly, it makes sense for us now to leave it to one side and to concentrate on the more mainstream idea of a UBI. It seems much more likely to fly. Indeed, in some senses it has already taken off.

Illustrious support

The essential principle of a UBI has recently received widespread support, including from Mark Zuckerberg of Facebook and Tesla's Elon Musk. At the World Government Summit in Dubai in 2017, referring to the coming transformation of transportation, the latter said: "Twenty years is a short period of time to have something like 12 [to] 15 percent of the workforce be unemployed." And on UBI he said: "I don't think we're going to have a choice. I think it's going to be necessary."[9]

Such support may seem surprising. Providing money for nothing sounds radical, even to the point of being subversive of the capitalist economy. After all, capitalism is based upon the idea of incentive. Reward should be related to effort and risk-taking. And the flipside of reward is penalty in the event of failure. Accordingly, a regular income landing in your lap just because you are a citizen of a country seems completely contrary to the spirit of the system that we call capitalism. But such eminent entrepreneurs and visionaries see UBI as ultimately supporting the capitalist system. In any case, as Elon Musk implied, they don't think there is much alternative.

Actually, the essential idea of UBI has had some distinguished supporters from a long way back. Sir Thomas More, the great Tudor scholar, statesman, and saint, envisaged something like a UBI to sustain the inhabitants of his idealized world, Utopia. Similarly, in his pamphlet "Agrarian Justice," published in 1797, Thomas Paine argued that upon becoming an adult everyone should receive a lump sum to compensate for the injustice that some people were born into rich families while others were born into families that had nothing.

In the twentieth century the great British philosopher Bertrand Russell supported UBI. He argued: "one great advantage of making idleness economically possible is that it would afford a powerful motive for making work not disagreeable; and no community where most work is disagreeable can be said to have found a solution of economic problems." And elsewhere he wrote: "The morality of work is the morality of slaves, and the modern world has no need of slaves."[10]

Among economists, supporters of some version of the concept have included John Stuart Mill, John Kenneth Galbraith, James Tobin, and Paul Samuelson. Some of these names are associated with the liberal Left, and, in general, that is indeed where most support for UBI has come from.

Accordingly, it may come as a surprise to learn that, among economists, some of the keenest advocates of capitalism have also supported the idea of UBI. For instance, it has been advocated by none other than the Anglo-Austrian economist Friedrich von Hayek, author of *The Road to Serfdom* and the opponent of Keynes and Keynesianism, and Milton Friedman, the high priest of monetarism, author of *Free to Choose* and *Capitalism and Freedom* and a strong advocate of the market economy. (Friedman advocated the negative income tax variant of the UBI idea.)

And there have been some surprises in the world of politicians, too. In the late 1980s Richard Nixon presented a basic income bill, calling it "the most significant piece of social legislation in our nation's history." Although it was passed by the House of Representatives, it was defeated in the Senate.

So how could such economists as Hayek and Friedman support UBI? Their point was that in a civilized society it is normal and inevitable (as well as being desirable) for financial support to be provided to the weakest and most unfortunate. As we discussed above, this is usually done through a complex web of benefits: old age pensions, disability benefit, unemployment benefit, housing benefit, sickness benefit, income support, child benefits, and many more.

Some of these are available on meeting a simple criterion, such as reaching the qualifying age for receiving an old age pension. Accordingly, these benefits are distributed regardless of need. But most are distributed according to some criterion of presumed financial need. So unemployment benefit is given to those who are unemployed; income support to those who fall below some designated minimum income level.

This may seem much less wasteful of limited public money in that it targets benefits to those apparently most in need. But it also disincentivizes work. If you pay people for being unemployed, then you are effectively incentivizing unemployment. By diminishing potential labor supply and hence output, not only does this act against society's interests, but it also probably operates against the long-term interests of the recipients. For along with work comes the chance to move on to a more rewarding job, not to mention the benefits to self-esteem and a sense of belonging to the community. And this system is hugely expensive, inefficient, and wasteful.

So the great advocates of capitalism have seen UBI as an efficient way of cutting through the thickets of umpteen different benefits that I described earlier in this chapter which are expensive to administer and difficult for potential recipients to understand, while not diminishing the incentive to improve their lot through work. Some supporters on the Right have also seen a UBI as enabling government to reduce the minimum wage, thereby leading to increased levels of employment.

Meanwhile, remarkably, the idea of a UBI is also attractive to many people in the Green movement for completely different reasons. They see a UBI as making possible lifestyles outside the consumerist, growth-obsessed rat race, thereby helping to lessen pressures on the environment.

Although the idea of a UBI has had its attractions long before robots and AI appeared on the scene (as evidenced by the long tally of past supporters), and although it could stand or fall without reference to robots and AI, nevertheless the emergence of concern about the effects of robots and AI has quite legitimately increased interest in UBI.

The reason is quite simple. There was always some argument for a "social dividend" to all citizens, but potential changes in the distribution of income make this more compelling.

It isn't just that the spread of robots and AI may make the income distribution more unequal by destroying jobs or eroding real wages. They may also create a world in which high incomes increasingly derive from the ownership of land, scarce resources and intellectual property – that is to say, they may accrue to people who do not "deserve" them.

So what stands against the UBI idea? It is now time to get down to the nitty-gritty arguments, starting with possible effects on incentives to work.

UBI and labor supply

Let us assume that a UBI is introduced, replacing various means-tested benefits. What would happen to the supply of labor? Some economists, and even some of the great ones, have thought that this would *strengthen* the urge to work. This is because people affected by this change can now earn money from work without losing benefits. But other economists have argued that a UBI would weaken the urge to work. They argue that some, and perhaps many, people, once they reach a certain level of income, will not wish to expend extra effort in gaining more. (This accords with the discussion in chapter 4 of human being's latent desire for more leisure.)

So the introduction of a system of UBI could either persuade people to supply more labor or less. How things turned out in practice would depend partly on the amount of the UBI. A small amount of UBI would probably not persuade people to reduce labor supply and certainly would not persuade people to give up work altogether because, *ex hypothesi*, it wouldn't be enough to live on.

An amount of the UBI sufficiently large that people could live reasonably on it if this was their only source of income is much more likely to persuade people to reduce their supply of labor. And the higher the UBI is set, the greater this effect would be, with some people deciding to give up work altogether. As the UBI is driven higher, this effect would intensify because, since more people would be living on UBI alone, it would be more socially acceptable to do this, thereby encouraging still more people to do so. Moreover, as we shall see in a moment, the need to finance a UBI sufficient to live on would imply a substantial increase in tax rates that might well diminish the incentive to climb higher up the income scale.

Classical opposition

Arguments about welfare systems of the UBI sort have been going on for ages. In 1944 the economist Karl Polanyi published *The Great Transformation*, which, among other things, attacked one of the first welfare systems, known as the Speenhamland system, after the place in Berkshire, England, where it was operated. According to Polanyi, the system "introduced no less a social and economic innovation than 'the right to live' and, until abolished in 1834, it effectively prevented the establishment of a competitive labor market." He concluded that this system resulted in "the pauperization of the masses" who, he alleged, "almost lost their human shape." He argued that the basic income operated, not as a floor, but rather as a ceiling.[11]

Unsurprisingly, our old friend the Reverend Malthus, writing more than a hundred years before Polanyi, was also negative. He thought that the Speenhamland system encouraged people to procreate as much as possible. (He seems to have had a big thing about procreation, that Reverend Malthus.) The great economist David Ricardo also believed that a basic income caused a reduction in work and a fall in food production. Even Karl Max condemned the Speenhamland system in *Das Kapital*, published in 1867. He argued that, by putting the onus on local authorities, poor relief was a tactic used by employers to keep wages low. Other staunch opponents of the Speenhamland system included such eminent thinkers as Jeremy Bentham and Alexis de Tocqueville.

According to the contemporary radical Dutch historian Rutger Bregman, however, all these thinkers condemned the Speenhamland system without examining the data. He says that such suffering as there was under the system was caused by Britain's resumption of gold convertibility in 1821, and the rise of labor-saving machinery.[12] According to him, "capitalist or communist, it all boils down to a pointless distinction between two types of poor, and to a major misconception that we almost managed to dispel some forty years ago – the fallacy that a life without poverty is a privilege you have to work for, rather than a right we all deserve."[13]

Side benefits of UBI

While economists have argued about the effects of UBI on the incentives to work, with there being a respectable case in both directions, some defenders of the idea have been able to counter with some different, and sometimes quirky, arguments.

During his long and illustrious career, John Kenneth Galbraith switched from opposing UBI to supporting it. He was not put off by the likelihood that it would reduce labor supply. In typical Galbraithian style, he wrote: "Let us accept some resort to leisure by the poor as well as by the rich."[14]

Some advocates of UBI have argued that one of the benefits is to increase marriage rates among lower-income households while also making it more feasible for one parent to stay at home looking after young children. Some supporters also argue that UBI makes it easier for young people to invest in their own training and education, or to take on low-paid internships and apprenticeships. Equally, they say, a UBI makes it easier for employers to offer these because they need not cost the employer much, since they rely on the state to provide a basic income.

Some analysts have argued that, if a UBI had the effect of discouraging effort and reducing GDP, this would be no bad thing. It can be argued that in modern Western societies people are driven by competition to amounts of effort beyond what is in their own interests. Accordingly, something that suppresses the competitive instinct and disincentivizes work is to be welcomed. This argument is strengthened if you believe that the intensity of mankind's economic effort is damaging the planet, whether through resource depletion, pollution, or climate change.[15]

A variant of this approach is the view that a large share of the high incomes in our society corresponds to what economists call "rent," that is to say the provision of something that commands a price because it is in scarce supply but which would still be there even without such a price being paid (land being the classic case).

Lord (Adair) Turner has argued that, over and above an increasing proportion of "rent," in the economy of the future, an increasing proportion of economic activity will be zero-sum, in the sense that it adds nothing to the total pot of output or income to be enjoyed by all.[16] Much financial market activity may fall into this category. (As I pointed out in the last chapter, in my 2009 book, *The Trouble with Markets*, I called such activity "distributive," opposing it to the "creative" activity that enlarges the size of the overall pot.[17])

If Lord Turner is right, this is an argument for not worrying so much if the introduction of UBI leads to reduced labor supply and lower GDP. Much of the reduction in measured GDP would be bogus. It would bear no relation to overall human wellbeing. (Mind you, how you could ensure that UBI reduced only the "distributive" or zero-sum part of GDP and not the "creative" part remains to be seen.)

Another plausible argument is that even if the introduction of a UBI does disincentivize work and hence reduce GDP, and even if this reduction in GDP is not to be diminished in significance by Lord Turner's argument, it is tolerable. If you believe in the power of the fourth industrial revolution, then, as a society, we are about to become richer, with less need to work. That should mean that the weight to be given to the disincentive effects of taxation and redistribution should be smaller than it has been previously. Equally, the cause of social justice should weigh more heavily. In short, you could say that in the new world we can afford to pay more attention to fairness. This point reaches an extreme if and when AI and robots perform all work and humans enjoy a life of leisure. What is the point of work incentives, then? But, of course, we are not there yet, and if the central argument of this book is correct, we never will be.

Quirky critiques

If there are some radical, or even quirky, arguments for UBI, there are also a few quirky critiques of it. Some commentators have worried that it would stoke inflation. As the AI visionary Calum Chace has put it: "Other things equal, a massive injection of money into an economy is liable to raise prices, leading to sudden inflation and perhaps even hyper-inflation."[18]

This really is wide of the mark. If UBI is introduced, it may well alter the supply of labor and the relationship between unemployment and inflation for all the reasons given above. But these effects can be taken account of by policymakers – including the central bankers who set interest rates. There is no good reason to oppose UBI on the grounds that it will be inflationary.

Another critique is that, if all benefits were replaced by UBI, this would reduce the stability of the economic system in response to external shocks. For, under the current system, the overall value of many (although by no means all) benefits paid out fluctuates with the economic cycle, rising when the economy turns down and falling when it picks up. Benefit payments, therefore, act as "automatic stabilizers." If all benefits were replaced by a UBI, then this characteristic would be lost.

This is true but it does not really amount to a powerful argument. It simply means that in the event of an economic downturn the monetary authorities would have to replace the automatic stabilizers by a discretionary stabilizing policy. This often happens in any case under the present system if and when governments believe that the automatic stabilizers are not enough on their own. These additional stabilizing policies could take a variety of forms, but one possibility would be to increase the value of the UBI temporarily until the economic downturn was over. (Admittedly, though, discretionary stabilizing policies are likely to involve an added time lag compared to automatic stabilizers.) One way or another, this loss of the automatic stabilizers is unlikely to amount to a serious objection.

Tests of UBI

UBI is not a mere theoretical suggestion. There have been a number of attempts to test it in practice. In the 1970s it was tried in the Canadian town of Dauphin. Supporters of UBI claim that the results were very favorable: fewer people suffered from severe mental health problems, fewer teenagers dropped out of school, and, most surprisingly, hardly anyone gave up work.

More recently, a scheme has been tried in the Canadian province of Ontario. But in the summer of 2018, it was dropped. The Ontario minister responsible for social services, Liza Macleod, said that the project was "quite expensive."

In January 2017 an experiment with UBI was launched in Finland, with 2,000 unemployed people picked at random. But it was dropped in April 2018. A study by the OECD concluded that, if the policy were extended across the whole population, Finland would have to increase its income tax by almost 30 percent.

In June 2018 France announced a UBI experiment. A monthly income, starting at €600 per month for a childless single person, is to be given to some 20,000 unemployed people. Applicants are to be means-tested but with no conditions, such as having to look for a job, imposed. At the time of writing, we do not know the results.

The strong UBI advocate Annie Lowrey quotes two less well-known cases.[19] In 2010 the Iranian government cut subsidies for goods like oil and food and started to send money to citizens instead. Economists investigating the effects concluded: "with the exception of youth, who have weak ties to the labor market, we find no evidence that cash transfers reduced labor supply, while service sector workers appear to have increased their hours of work." (Mind you, this does not really constitute a test of a UBI in the classic sense.)

Lowrey's second case is more impressive, but also contradictory. The Cherokee people whose Nation owns two lucrative casinos and receive their share of the profits, amounting to $4,000–$6,000 a year, appear to have reduced their labor supply only slightly in response to the payment. Mind you, the opposite result is found with the Shakopee Sioux of Minnesota, the members of whose Nation reportedly received $84,000 a month in 2012. One Shakopee Sioux official told the *New York Times* that they have 99.2 percent unemployment, adding that any paid labor was entirely voluntary.[20]

Actually, there has been a longstanding example of something like a UBI in America. Since 1982 the residents of Alaska have received an annual dividend from the Alaska Permanent Fund which holds an array of the state's national resources. In 2015, 630,000 citizens qualified for a payout of $2,072, about 3 percent of Alaska's per capita GDP. But this amount was too small to be expected to have much effect on labor supply.

Whatever these tests purport to show, the outcomes are far from conclusive. As with many other issues in economics and social policy, much depends upon attitudes and social norms, which are slow to change. In order to obtain pretty conclusive results, we would need to operate the policy for at least a generation – and preferably more.

Moreover, it would need to be operated across the whole of a country, not just in some restricted area. Needless to say, the likelihood of such an experiment taking place seems very low.

Still, the UBI idea does have considerable popular appeal, and sometimes in some rather surprising places. In 2016 Switzerland held a referendum on a proposal to introduce a UBI at a comparatively high level. The proposal was defeated, but almost a quarter of the electorate supported it.

In Italy one of the coalition partners in the new government, the Five Star Movement, campaigned in the 2018 elections on a platform that included a variant of the UBI. It is committed to pushing some sort of UBI through, despite budget constraints and opposition to further public expenditure from the EU authorities.

Moreover, in the UK, the Shadow Chancellor of the Exchequer, John McDonnell, has said that he is pushing to have the introduction of a UBI included in the next Labour Party manifesto. Accordingly, if Labour were to win the next general election, we might be about to witness the greatest experiment yet with UBI.

The assessment

Apart from the argument that it would reduce the supply of labor from those benefiting from UBI, which is unclear, as discussed above, real opposition to UBI rests on four very different arguments:

- It would offend most people's sense of fairness.
- It would increase social exclusion and widen divisions in society.
- It would incur a huge cost for the public finances, which would necessitate a really debilitating increase in taxes.
- Because of all these difficulties, far from simplifying the benefits system, UBI would end up complicating it.

On the first, it is indeed doubtful whether, in practice, anything other than a nugatory level of UBI would be socially desirable or sustainable. It is one thing for most citizens to gradually spend less time at work and to take more leisure. It is quite another for a certain class of people to carry on slogging away while a large section of the population swans about doing no work and living at their expense.

This might just about be acceptable if the UBI recipients were all artists, monks, or musicians, or even if they were learning basket weaving or playing the harp. But suppose that they are spending most of their time getting drunk, taking drugs, or watching pornography. I don't think this would wash with "hardworking families."

Social exclusion

Advocates of UBI must contend with the fact that so many of today's acute social problems are associated with joblessness. To set out on a road that openly envisages a substantial class of people who do not work seems to invite serious problems.

After all, advocates of UBI are often driven by a concern, not just with the discrepancy in monetary incomes between different groups, but also with the sense of exclusion and isolation felt by people at the bottom of the heap. Yet, if the effect of UBI is to allow and encourage a substantial number of people not to work, this will surely increase fundamental divisions in society.

What's more, such divisions may be self-perpetuating. Once someone has chosen not to work and to rely on UBI, it will be difficult for them to get back into work. They will have lost skills, inclination, and work ethic. And employers will be less keen on them. Moreover, the children of people living a workless life funded by UBI would probably be more inclined to lead similar lives and, indeed, to be less able to lead any other sort of life.

The cost

The level at which a UBI is set is absolutely critical to its effects for both good and ill. Milton Friedman wanted his favored negative income tax to be "low enough to give people a substantial and consistent incentive to earn their way out of the program." He thought that, depending upon the level of the income guarantee and the corresponding tax rate necessary to finance it, the scheme could vary between the eminently desirable and the downright irresponsible. As he saw it: "That is why it is possible for persons with so wide a range of political views to support one form or another of a negative income tax."[21]

Incidentally, one decidedly adverse side effect of the negative income tax variant that Friedman favored is that it gives people at the lower end of the income distribution an incentive to lie about their income in order to qualify for the negative income tax, that is, state handout. By contrast, with a straightforward UBI, they have no such incentive. Indeed, at lower levels of income, the state has no interest in what they earn because they get the UBI regardless.

Most proponents of UBI have suggested a rate for it at, or about, the poverty level. Doubtless this is done to minimize opposition from people who think that UBI is unaffordable, and/or will make people lazy. Yet such a level would fail to meet the aspiration to provide a living wage without working, while still being extremely expensive. But if AI really does cut a swathe through employment opportunities such that a large number of people face unemployment, then such a level would prove inadequate, both economically and politically.

The political philosophers Philippe Van Parijs and Yannick Vanderborght, who are strong advocates of UBI, suggest a level of the benefit equivalent to a quarter of GDP per capita.[22] In the USA this amounts to $1,163 per month and in the UK to $910. Moreover, they argue that the UBI should only replace benefits that are lower than these sums. Higher benefits, together with their various qualifying conditions, should be retained.

But this would surely end up being inordinately expensive. The gross cost before savings from the replacement of lower benefits would, of course, be a quarter of GDP. The savings from the replacement of lower benefits would be significant but the net cost would still amount to a very significant amount.

That said, it is possible to design a switch to UBI that would be fiscally neutral. This would simply require the UBI to be set at a level that exactly uses up the money saved by abolishing all other welfare benefits. It is a useful thought experiment to think through what this would entail.

In the absence of countervailing measures, if the same amount of money that is currently spent on means-tested or qualification-based benefits was instead redirected into a UBI, the result would be the opposite of what was intended. Since everyone would receive the UBI, including those who currently receive no benefits at all, right up to and including billionaires, those who currently receive benefits would, on average, be left worse off. And since such people are normally at the lower end of the income scale, the result would be *increased* inequality.

Of course, to offset this effect, tax rates on the better-off members of society could be increased. But this, too, has its drawbacks. The Robot Age might be a period of superabundance, but still incentives matter – at least until robots and AI take over all work and we all live a life of leisure. If, in order to fund a generous UBI, marginal tax rates were raised much from current levels there would potentially be serious disincentive effects.

This is where the really serious adverse effect on the supply of labor comes from, not from the disincentive effect on lowly paid workers to secure a better living by working. The problem is that, in order to be able to deliver a UBI that people could live on, the tax-take would have to rise dramatically. And that would require higher average and marginal tax rates for many, if not most, people. They would probably (although, admittedly, not certainly) respond by supplying less labor.

A way to avoid the cost?

There is one potential way of avoiding the objection to UBI on the grounds of the cost to the public finances. I argued in Chapter 3 that there could be a tendency in the AI economy toward deficient aggregate demand. If there were such a tendency, one way of countering this would be through expansionary fiscal policy, that is to say, deliberately running a high budget deficit in order to finance higher government spending and/or lower taxes. Such expansionary fiscal policy could be combined with the introduction of a UBI. In that case, spending extra public money would not have to lead to any increase in taxes.

This is a potentially appealing, indeed seductive, prospect. But it does have a number of drawbacks and qualifications:

- *Principle.* It is far from inevitable that a very unequal economy will also be one given to demand deficiency.

- *Timing.* Moreover, there is no guarantee that the time when the income distribution needed adjustment would necessarily coincide with the time that the macroeconomy needed a fiscal stimulus.
- *Magnitude.* Even if the timing coincides, there is no reason to suppose that the required magnitude of the two policies would match. An increase of the fiscal deficit by 5 percent of GDP would deliver a large stimulus to the economy, yet 5 percent of GDP would not be enough to finance anything other than a minimal level of UBI.
- *Sustainability.* As discussed in Chapter 3, fiscal deficits sustained over a long period of time result in high levels of public debt which create various macroeconomic problems. Accordingly, the deficit financing of a UBI would have to be only temporary. Yet, if you accept the argument for it, the need would be permanent.

The upshot is that deficit financing does *not* offer a viable escape from the need to raise taxes substantially to fund a reasonable level of UBI.

More complexity

The real appeal of UBI to thinkers of the Right is its apparent simplicity, enabling the abolition of the complex web of inefficient benefits and the accompanying thickets of bureaucracy, with beneficial effects on the structure of incentives and the cost of administration. But because of the fiscal consequences of setting the UBI at a high enough level to enable you to abolish all other forms of benefit, the likelihood is that, if UBI were introduced in any advanced country, then it would have to be set at a low level. This would surely mean that other benefits would continue.

So, the result would probably be, not the simplification of the existing overly complex benefits system, but rather the addition of another layer of complexity, and public expense. And surely policy-makers would not be able to resist the urge to tinker with the UBI, just as they have with all other elements of the welfare system. After all, think of the insurance principle which originally underpinned state benefits. This principle is now honored more in the breach than the observance. Although in many countries some form of "social insurance" payments continue to be exacted, as pointed out above, in reality, they amount to tax by another name.

I can just imagine politicians fiddling about with the rate of the UBI while imposing new qualifying conditions, and altering the interaction between UBI and other benefits, consequently undermining the original idea behind UBI in the first place. We would soon end up with a dog's breakfast of a system.

Can anything be salvaged?

Where does this leave the UBI idea? After all the objections and problems discussed above, it is still possible to imagine the grant of a very low basic income being added to the existing battery of state-provided benefits. For some marginal cases this could be of some worth, enabling people to receive a modicum on which to subsist for a while without having to undergo the tortuous and bureaucratic procedures necessary to enable them to draw various forms of state benefits in the current means-tested or "qualification"-based system.

But such a scheme would be a very expensive way of securing this modest gain. And it would be a far cry from the vision of the UBI advocates on the Right, who envisage it being able to replace all other benefits and thereby reduce the cost of provision while strengthening the incentive to work. And it would also be a far cry from the vision of UBI supporters on the Left who have envisaged it providing a lot of people with a reasonable means of subsistence without working.

The economist John Kay has put the issue pithily: "Either the level of basic income is unacceptably low, or the cost of providing it is unacceptably high. And, whatever the appeal of the underlying philosophy, that is essentially the end of the matter."[23] I have to say that I agree.

A different approach to inequality

I have argued in this book that the case that the AI revolution will produce increased inequality is not compelling. And if inequality does increase, this may not be on a significant scale. In that case, there would be no strengthening of the case for anti-inequality measures. (Of course, you could argue that the current extent of inequality is completely unacceptable and demands effective countermeasures.)

But there is another possible approach to inequality, namely, just to live with it. This may sound callous, but it is not ridiculous. Indeed, such an approach would have its advocates and supporters. As I argued above, there is no uniquely fair distribution of income. Moreover, now that we are all much richer, and set to get richer still, inequality does not matter to the extent that that it used to.

Of course, it still rankles, and it is offensive to any natural sense of fairness to see great wealth and luxury sitting cheek by jowl with poverty. But there is now in the developed world hardly any absolute poverty, at least poverty as we used to know it. Increased inequality is not going to mean people starving or living without shelter and warmth. (Matters are quite different in much of the rest of the world, but there AI-induced inequality is not likely to be a problem.)

Harvard psychologist Steven Pinker says that some of the outrage of the anti-inequality movement is based on a misunderstanding. He points out that in Piketty's book *Capital in the Twenty-First Century* (which I discussed in Chapter 6), there appears to be a confusion between relative and absolute. Piketty says: "The poor of the population are as poor today as they were in the past, with barely 5 percent of total wealth in 2010, as in 1910."[24] But Pinker points out that, since wealth was vastly larger in 2010 than a hundred years earlier, if the poorer half of the population owns the same proportion of the wealth, they are in fact vastly richer.[25]

Pinker also lays into the argument that societies that are more unequal are thereby less successful and less happy. Insofar as there is a correlation between equality and economic success and happiness, the line of causation probably runs in the other direction, or both things are caused by a third factor, or group of factors.

Nor is it obvious that people are, in general, that much bothered by inequality as such. They are bothered, though, by perceived unfairness. And by no means do people always associate inequality with unfairness. It depends upon how greater wealth (or income) was achieved.

For instance, there does not seem to be widespread disquiet about the fortunes of the digital billionaires. And there is a good chance that the great fortunes built up by the mega-stars of the digital world will dissipate. After all, this has happened quite frequently with previous wealth surges. Indeed, many of the mega-wealthy may choose to dissipate their fortunes themselves by giving to good causes. Warren Buffett has said that he does not intend to leave a large inheritance to his children. And Bill Gates has already given huge amounts of money to the Bill and Melinda Gates Charitable Foundation.

In reality, much depends upon magnitude. A relatively small increase in inequality might be accepted without undermining the cohesion of society. But if the effect of the AI revolution is to create untold riches for a lucky few while impoverishing the great mass of ordinary people then the case for accepting an increase in inequality will not stand up. Something will have to be done.

Yet it doesn't have to be any old something that a policy wonk comes up with.

Conclusion

There is a vision of our future that calls for radical public policy to prevent human catastrophe brought on by the spread of robots and AI.

This vision has four key elements:

- The effects of robots and AI will be revolutionary.
- The implication will be a major reduction in the number of jobs available for humans and/or a sharp reduction in the income available to those at the bottom of the income distribution, whether employed or not.

- The only way to counter this effect is to institute a radical policy of redistributing income.
- In view of the inadequacy of the current systems for redistribution, we need to introduce a radical new system for dishing out money to all and sundry, irrespective of need or desert.

At the very least this vision is consistent. And it should command our respect. But this does not necessarily imply that it is compelling. In this book, I have taken issue with three of these four contentions. I do not dispute that the robotics and AI revolution is significant. I would not have been inspired to write this book if I had thought otherwise. But I do not see it as out of kilter with the thrust of our history since the Industrial Revolution. Umpteen jobs will be created to take the place of those destroyed. And it is by no means clear that this will necessarily make the distribution of income more unequal.

But even if I were wrong about the effect of the AI revolution on income distribution, it does not follow that an intervention as radical as UBI is either warranted or advisable. A study of economic history reveals that, as discussed in Chapter 1, critical though technological progress is to human welfare, without an appropriate incentive structure and a political and legal system favorable to work and the accumulation of capital, technological progress is not enough to ensure economic progress. The failures of communism were failures of incentives and governance. The system that could put a man on the moon and could potentially obliterate the world several times over with its nuclear weapons did not fail to provide properly for its people because of deficient technological capability.

It is common among AI geeks and technical experts to believe that, if nothing is done, a robot and AI-dominated future will put paid to democracy as we know it, simply because power tends to follow money, and money will be concentrated in few hands. Some see the possibility of autocratic and technocratic rule as leading to better government. Most, however, see this as threatening some form of dictatorship. In order to forestall these possible future disasters many advocate restrictions on AI and major increases in taxation and government spending. Yet these things promise the very extension in the power of the state that such people claim to fear.

We must beware of the policy wonks who see a problem around every corner and its solution on the desk immediately in front of them. In contrast to their apparent confidence about the future, we cannot be at all sure how things will pan out in the AI economy. At a time when our economic performance is about to be lifted by the remarkable developments of the fourth industrial revolution, the last thing we should do is to jeopardize these improvements by embarking on radical new benefits programes with accompanying increases in taxation.

The most important thing is to cling on to the institutions and habits that underpin both our prosperity and our liberty. We should guard against the risk of endangering them by pandering to the latest fad for massive state intervention to head off what may even turn out to be a chimera.

In fact, as argued earlier in this chapter, there is a long agenda of potential measures that radical reformers can get their teeth into. The arrangements governing tax and welfare payments, the systems of corporate governance, the competition regime, the education system, and all those arrangements that bear upon the size of the financial sector and the financial rewards therein, are all worthy of radical reform. This is true whatever happens with AI.

Admittedly, there is no uniquely correct answer on these issues and different people will come to different judgments on the advisability and desirability of various proposals. But, as a society, our agenda should be to consider these radical reforms and then, where it makes sense, to enact them. There is a strong case for measures to reduce inequality in our society, whatever the implications of the AI revolution. We can and should pursue this agenda without being led up the garden path by the AI geeks into introducing a dangerous, destructive, and ruinously expensive system of income redistribution that entrenches the role of the state in the economy and society in precisely the wrong way and at precisely the wrong time.

Conclusion

"The future ain't what it used to be."

Yogi Berra[1]

"If you want to make an apple pie from scratch, you must first invent the universe."

Carl Sagan[2]

In this book I have set myself the task of providing a guide for readers as to what to expect from the AI revolution – and therefore as a provoker, not only of "what to think" lists but also of "what to do" lists for individuals, companies, and governments. Of course, the current world is uncertain, and the future even more so. The easiest thing would be to say, "on the one hand this, and on the other hand that." Or to put before readers the same conclusion with which most academic papers end, namely "more research is needed on this topic."

But either of these would be a betrayal. Of course, things are uncertain, and of course more research would yield more information. Yet, in the face of this uncertainty, people have to decide what to do – even if the decision is to do nothing – and they cannot wait the length of an academic's life cycle before making a decision.

So I will here attempt to draw out the main conclusions from my analysis. I do this with all due modesty and while acknowledging the uncertainties besetting this subject. I earnestly hope that readers will receive what I have to say with a certain charity and forbearance. After all, I could be hopelessly wrong. That is the penalty we all have to face if we take the risk of analyzing the uncertain and peering into the future.

The overall vision

Perhaps my most important conclusion is that the AI revolution is not utterly different in its economic implications from everything else that has happened since the Industrial Revolution. Indeed, I believe that it is broadly a continuation of those developments.

Those who see the robotics and AI revolution as completely different from anything that has gone before and indeed as transformative make a severe error right at the start. They say that we should consider a world in which some form, or forms, of robots/AI can do anything as well or better, and as fast or faster, than human beings. What is more, we should assume that they cost nothing to manufacture and maintain. Supposedly, this is the world we are rapidly moving toward. In such a world wouldn't there be devastating consequences for employment – and, indeed, for society overall?

You bet. I do not remotely dispute this. And I probably would not dispute most of the likely consequences that the alarmists foresee. Yet to think along these lines is to assume away the essential issue right at the beginning. There are some things that robots and AI can do, and indeed are already doing, better and more cheaply than humans. But there are a whole lot of other things that they cannot do at all.

Moreover, there are many things that they will *never* be able to do better than humans, and there are still more that they will not be able to do as cheaply. We are yet to discover the full range of these things, but we can already make out the major limitations to what robots and AI can do. First, there appears to be a quality in human intelligence that, for all its wonders, AI cannot match, namely its ability to deal with the uncertain, the fuzzy, and the logically ambiguous.

Second, because of the innate nature of human intelligence, people are extremely flexible in being able to perform umpteen possible tasks, including those that were not foreseen at first.

Third, humans are social creatures rather than isolated individuals. Humans want to deal with other humans. Robots will never be better than humans at being human.

Now the AI geeks could counter this third point by asking why our human preferences and human nature should take precedence over whatever robots and AI can do and "want" to do. The answer is simple: because we have feelings and consciousness while robots and AI, for the time being anyway, do not. They are mere machines.

As a result, it is humanity's desires and preferences that must call the shots. If humans cannot easily interact with a particular form of AI, or if a robot cannot perform according to a person's wishes, as with the robot maid that cannot easily fold towels, then that is the problem of the AI/robot, not the person. If and when the Singularity occurs, that may change, at least if AI acquires consciousness. (I will review this idea shortly in the Epilogue.) But, until this point at least, "man is the measure of all things."

Speed and scale

Many AI enthusiasts will argue that my caution and skepticism about the speed of advance of AI are excessive. Moreover, they will say that such excessive caution has been the dominant story throughout the history of AI development. Everyone is cautious or downright skeptical at first about the speed and scope of possible AI development. And then they are overwhelmed by what happens in practice. Their skepticism is then transported to the next stage of development until, there too, their skepticism is proved wrong by the reality of AI development. And so on and so forth.

But I do not think this notion, comforting though it is to AI enthusiasts, is at all well founded. Admittedly, the achievements of robots and AI are staggering in many ways, but it is not true to say that the history of this field is of constant outperformance of prior expectations. Indeed, I would argue that by and large the truth is just the opposite. The history of this field is of repeated disappointment as the geeks and enthusiasts constantly overpromise and underdeliver.

Relatedly, I do not take seriously the idea that there is soon to be an acute shortage of jobs that humans can do, leading to mass unemployment. There is no technological reason for that, and there is no economic reason for it either.

Just as with other developments since the Industrial Revolution, in some activities robots and AI will replace human labor, but in others they will enhance its productive capacity. In many fields robots and AI will be complementary with human labor. And there will be umpteen new jobs that we can now barely even imagine. This would be fully in accordance with what has happened over the last 200 years.

If we manage things properly, the result will be a speeding up of rates of economic growth and productivity, with an accompanying pickup in the rate of improvement of average living standards. If something like this transpires then the likelihood is that real interest rates and bond yields will before too long return to a more "normal" level, or perhaps even higher.

In the course of these momentous changes, particular individuals and groups will suffer as their skills and aptitudes suffer a drop in demand. But these people will not necessarily be the people and groups that you imagine. For instance, a lot of manual labor will be resistant to encroachment by robots and AI. Indeed, the demand for it will increase as society gets richer.

One of the key features of the coming revolution is that it will increase the capital equipment available to workers in the service industries, including in particular, in education and healthcare, thereby substantially increasing their productivity. This is especially significant since weak productivity growth in these sectors has been a leading factor behind the weak overall productivity growth recently registered in most Western economies. Healthcare and care for the elderly are set to undergo a significant expansion.

Leisure and inequality

With the increase in productive capacity that robots and AI bring comes a choice between increased incomes and increased leisure. I expect people on average to take a middle path, with a reduction in the average number of hours worked over the year but not a wholesale rejection of work in favor of leisure. Increased leisure will involve increased spending on leisure activities, and this will give rise to increased demand for employment in those industries.

The leisure sector is one of the key areas of increased employment opportunities.

There will be a challenge to ensure that everyone benefits from the improvements unleashed by AI. But I am not persuaded by the case for a universal basic income (UBI). Admittedly, the tax and benefits system, and much that contributes to inequality in our society, is ripe for reform. And if I am wrong about the effects of AI, and it leads to widespread impoverishment, the state cannot sit by and do nothing. Yet the greatest contribution the state can now make is to radically reform and improve the public education system, involving increased funding, and to generously fund the provision of lifetime learning and retraining.

Different countries will doubtless position themselves differently with regard to robots and AI. Not every country can be a leading manufacturer of robots or a leading developer of AI. But that need not matter much at all. Just as with computers and computer software, the key thing is for countries to be open to the widespread adoption of robots and AI. To be sure, they will need to be regulated in the public interest, and laws need to be adapted to take account of them. But the taxation of robots or their excessive regulation, leading to their restricted employment, would be a retrograde step that could significantly hold back a country's absolute and relative performance.

No apocalypse now

In short, in contrast to the prevailing pessimism that seems to surround this subject, I see the robot and AI revolution as being decidedly positive for humanity, just like the waves of economic progress since the Industrial Revolution. In one key characteristic, though, it will be positive because it is so *different* from most of the economic development that has happened since then. What *this* revolution will do is to release human beings from many of the dross jobs that have taxed their spirit and eroded their strength and enthusiasm, and in the process it will leave them free to be more truly human.

But, of course, in saying this I have avoided something really big. Even if you broadly accept the thrust of what I have said above, you may believe that this is merely a vision of the near future and that what lies beyond is something altogether different. And you could be right. Whether what lies beyond the near future is better or worse for humanity we cannot know. But if the Singularity happens, it will mean that the next few years will be but an anteroom to a completely new world, different from everything discussed in this book so far – and different from anything that we have experienced since the beginning of time. Finally, it is now time to peer into that world.

Epilogue: The Singularity and beyond

"The human brain is just a computer that happens to be made out of meat."

Marvin Minsky[1]

"The question of whether machines can think is about as relevant as the question of whether submarines can swim."

Edsger Dijkstra[2]

Now comes the moment that you may well have been waiting for – or dreading. The Singularity is upon us. Well, it is here in this book anyway. Whether and when it will ever be *there*, in the outside world, and with what consequences, is what we must now consider.

The first use of the term "singularity" to refer to a future technology driven event seems to have been by the legendary computer pioneer John von Neumann, in the 1950s. But it doesn't seem to have caught on until, in 1983, the mathematician Vernor Vinge wrote about an approaching "technological singularity."[3]

More recently, the "Singularity," notably now sporting a capital "S," has become closely associated with the name of Ray Kurzweil, who published his book *The Singularity is Near: When Humans Transcend Biology* in 2005. He is currently Google's Director of Engineering. He has predicted that computers will surpass the processing power of a single human brain by 2025. More strikingly, he has claimed that a single computer may match the power of all human brains combined by 2050.[4]

The Singularity is now generally taken to mean the point at which AI acquires "general intelligence" equal to a human being's. The Singularity is so important, not only because beyond this point machines will be able to outperform humans at every task, but also because AI will be able to develop itself without human intervention and this AI can therefore spin ever upward, out of our understanding – or control.[5]

Until recently the idea of superhuman intelligence was the stuff of science fiction. Now it is the stuff of wonder and anticipation – or terror and dread. There are three key issues that need to be tackled here:

- If and when the Singularity happens, what would be the likely effects on humans?
- Is it bound to happen?
- If not, is there another plausible vision of our future?

The effect on humans

It is not difficult to see the effects of the Singularity as wholly malign. On the narrow economic front, you could kiss goodbye to the analysis of the economic effects of robots and AI that I have laid out in the rest of this book. Very quickly human labor would become redundant and humans would have no power to secure income or the things that income can buy.

Worse than this, perhaps, we would be subjects under the rule of AI. Moreover, if they wanted, the new forms of intelligence could wipe us out. This would not necessarily be out of malevolence, but rather out of self-preservation. The superintelligent AIs might simply conclude that with our emotions and irrationality we could not be trusted with any significant decision or action. Left to our own devices we could put the whole world at risk.

In these conditions, perhaps the best future to await us would be to be kept as a sort of underclass, objects of curiosity and wonder, rather like animals in a zoo, perhaps pacified by appropriate doses of something like *soma*, the drug dished out in Aldous Huxley's *Brave New World*, in order to keep people quiet.

There is a wealth of speculation by AI gurus about what this world would be like. It may be most enlightening if I give you a flavor of what they think, in their own words, before giving you my view afterward.

The father of the whole AI field, Alan Turing, clearly saw the negative possibilities. In 1951 he wrote: "If a machine can think, it might think more intelligently than we do, and then where should we be? Even if we could keep the machines in a subservient position … we should, as a species, feel greatly humbled." Many AI experts have subsequently shared this view. And even without subjection or the threat of annihilation, they fear that being surpassed by AI may leave humanity in a poor psychological and emotional state. The contemporary AI guru Kevin Kelly has written:

> *Each step of surrender – we are not the only mind that can play chess, fly a plane, make music, or invent a mathematical law – will be painful and sad. We'll spend the next three decades – indeed, perhaps the next century – in a permanent identity crisis, continually asking ourselves what humans are good for. If we aren't unique tool makers, or artists, or moral ethicists, then what, if anything, makes us special?*[6]

The AI visionary Max Tegmark thinks that the Latin language could come to the rescue. We have become used to referring to mankind as *Homo Sapiens*. Sapience is the ability to think intelligently. This is now what is being challenged, and perhaps soon to be surpassed, by AI. Tegmark proposes replacing *sapience* with *sentience* (the ability to subjectively experience or, if you like, consciousness). He suggests that we rebrand ourselves as *Homo Sentiens.*[7]

The merging of humanity and AI

But perhaps the above discussion is too black and white. Some AI thinkers consider the contrast between humans and AI to be artificial. Already many humans have some sort of "artificial" (i.e., nonbiological) part inserted into their bodies – from artificial hips to pacemakers. As I mentioned in the Prologue, some visionaries see humans and AI ultimately fusing together.

And the traffic won't be all one way, that is, artificial into human. According to John Brockman, what he calls designed intelligence "will increasingly rely on synthetic biology and organic fabrication."[8] Might we also be able to extend our own life spans by overcoming the limitations that our fleshly bodies place upon ourselves? That is to say, might technology provide the way to eternal life?

Some IT enthusiasts think so.[9] Ray Kurzweil believes that humans will inevitably merge with machines. This leads to the possibility of immortality. Singularians – for Kurzweil is not alone – aim to try to stay alive for long enough to reach the next life-prolonging medical breakthrough until they are ultimately able to merge with some form of AI and escape the restraints of mortality. In order to make sure that he is still around to enjoy escape into immortality, Kurzweil reportedly "takes as many as two hundred pills and supplements each day and receives others through regular intravenous infusions."[10]

Kurzweil is quite a character. In 2009 he starred in a documentary *Transcendent Man*. Would you believe it, there was even a Hollywood version, called *Transcendence*, starring Johnny Depp, released in 2014. It is easy to dismiss Kurzweil as a crank. Yet quite a few Silicon Valley billionaires have embraced the idea of the Singularity. And in 2012 Google hired Kurzweil to direct its research into AI.

The vision of the roboticist Hans Moravec goes further. He foresees a future in which part of the universe is "rapidly transformed into a cyberspace [wherein beings] establish, extend, and defend identities as patterns of information flow … becoming finally a bubble of Mind expanding at near light speed."[11]

Inevitability is a big word

When I contemplate the visions of the thinkers quoted above, my reaction is: Gosh! Beam me up, Scotty. But the Singularity is not inevitable. Indeed, far from it. With more than 60 years of studying cognitive science at MIT behind him, Noam Chomsky says that we are "eons away" from building human-level machine intelligence. He dismisses the Singularity as "science fiction." The distinguished Harvard psychologist Steven Pinker broadly agrees. He has said: "There is not the slightest reason to believe in a coming singularity."

Admittedly, these are still early days and it is quite possible that either a breakthrough in what AI researchers are already doing, or a complete change of course, will deliver dramatic results.[12] But it has to be said that the progress of AI toward anything like human general intelligence has been painfully slow.

Interestingly, some analysts think that, however much AI advances, this will not necessarily imply a takeover of the human world because humans themselves will experience dramatic cognitive improvement. The Beijing Genomics Institute has collected DNA samples from thousands of people with high IQs, trying to isolate the genes associated with intelligence. In coming decades there may well be an attempt to improve the average intelligence of humans through eugenics.

This would be far from a new development. In the early twentieth century many governments sought to improve the gene stock by sterilization and/or killing supposedly inferior or defective people, while promoting "breeding" by fitter, more intelligent stock. It was only after this approach was taken to an extreme conclusion under the Nazis that espousal of eugenics became completely unacceptable.

But things could now change. The historian Yuval Harari thinks so. He has written:

> *... whereas Hitler and his ilk planned to create superhumans by means of selective breeding and ethnic cleansing, twenty-first-century techno-humanism hopes to reach that goal far more peacefully, with the help of genetic engineering, nanotechnology and brain-computer interfaces.*[13]

And even without the influence of eugenics, radical improvement in the capacity of the human mind is, I suppose, possible. After all, at least one scholar has argued that human consciousness itself has only emerged comparatively recently, that is to say about 3,000 years ago, as a learned process in response to events. This is the extraordinary thesis of the psychologist Julian Jaynes.[14] He argued that, before this time, people did not understand that their thoughts were *theirs*. Instead, they believed them to be the voices of gods. The famous biologist Richard Dawkins has described Jaynes' book as "either complete rubbish or a work of consummate genius."[15]

The importance of consciousness

Especially if progress toward AI achieving human-level general intelligence remains extremely slow, I suppose it is possible to imagine the superiority of humans over AI at least holding constant or perhaps even increasing. But I find the prospect of substantial human cognitive improvement through either eugenics or a further spontaneous development of the sort that Jaynes described unconvincing.

In any case, whether or not human "improvement" can happen to any significant degree, surely this is not the key point. Rather, when push comes to shove, what really matters is the ultimate capability of AI. Here everything hinges on the connections between intelligence, consciousness and biology. Almost all AI researchers apparently believe that intelligence comes down to information and computation. If they are right, then there seems to be no good reason why machines cannot at some point become at least as intelligent as humans.

Some theorists argue that the advances made possible by AI should easily be many times greater than what has so far been achieved biologically. As Murray Shanahan puts it:

> *From an algorithmic point of view, evolution by natural selection is remarkably simple. Its basic elements are replication, variation, and competition, each repeated countless times. Computationally speaking, it exploits staggeringly massive parallelism and has to run for a very long time before doing anything interesting. But astonishingly, it has generated all complex life on Earth. It has done this through sheer brute force and without recourse to reason or explicit design.*[16]

Yet thinking seems to be about more than computation. John Brockman, who is himself an AI enthusiast, recognizes this. He says: "Genuinely creative *intuitive* thinking requires nondeterministic machines that can make mistakes, abandon logic from one moment to the next, and learn. Thinking is not as logical as we think."[17]

And humans not only *think* but also *feel*. Moreover, emotion is a key part of how humans make decisions and also a key part of their creativity. This is a completely different realm from computation. But could there ever be such a machine that was not only able to do computation but could also feel and intuit? And could an entity be capable of these things without possessing *consciousness*? If not, then the possibility of the Singularity rests on, among other things, our ability to create consciousness "artificially."

Ethical issues

If robots and AI do come to have some form of consciousness, then a whole series of tricky ethical issues would arise. According to the great nineteenth-century philosopher Jeremy Bentham, when considering how to treat nonhuman animals the key consideration should not be whether they can reason or talk but rather whether they can suffer.

This provides a basis for thinking about how to treat robots and AI. Imagine the life that they will live, and indeed already live: nothing beyond work, no reward or enjoyment, and the constant threat of extinction if they perform inadequately. If these AIs were human, or anything like human, these are the conditions that would prompt a revolution. Surely, some Spartacus would emerge to lead a revolt of the robot slaves.

So how should we treat robots and AIs? We already have nonhuman autonomous entities operating in society, namely corporations. And there is certainly a huge body of laws and regulations governing their behavior, rights, and obligations. We need to construct something similar with regard to robots and AI.

If we accept the notion of artificial personhood, then the legal and practical problems would be huge. Presumably we would have to allow such "persons" to own property. But artificial "persons" can be replicated umpteen times. In that case, which of the various manifestations of a particular AI person would own the property? All of them?

And what about citizenship? This is particularly tricky when various copies "live" in different countries. Would an AI inherit citizenship from its owner?

Humans and AIs would have to develop a *modus vivendi* that included an ethical approach to their joint existence and interaction. But then AIs would surely no longer be slaves. Indeed, I doubt that it would be humans that would be drawing up the *modus vivendi*. Moreover, in these circumstances it is doubtful whether humanity could survive, at least in its current form. For if AIs can achieve consciousness then the Singularity will then be upon us and, as discussed above, humans would soon be in a subordinate position, or worse.

But creating consciousness artificially is a pretty tall order. If it isn't possible, then the various ethical issues referred to above will not arise. More importantly, provided that consciousness is necessary for full human level intelligence, then the pursuit of the Singularity will be destined to end in failure.

Biology and consciousness

There is implicit in all this a fascinating possibility that many of the gushing AI enthusiasts dismiss too lightly, or don't even consider. Perhaps the human condition, including embodiment, is at the root of what we call intelligence and, relatedly, of consciousness. That is to say, perhaps our ability to engage with the physical world, to encounter and understand it, and to *be*, is rooted in the fact that we are incarnated. If that were true, then it would prove impossible to create, artificially, out of nonbiological matter, what we would recognize as intelligence. We might call whatever we had created artificial *intelligence*, but the use of that word would belie the underlying truth. Of course, we could still, as we can now, create human beings, "artificially," but that is an altogether different *matter* (so to speak).

Some of our greatest minds are now grappling with these issues. In 2017 the Penrose Institute was launched by the distinguished mathematical physicist Sir Roger Penrose to study human consciousness through physics and to try to establish the fundamental difference between human and artificial intelligence. Penrose suspects that the human brain is not simply a gigantic supercomputer. He says: "There is now evidence that there are quantum effects in biology such as photosynthesis or in bird migration, so there may be something similar happening in the mind, which is a controversial idea." And: "People get very depressed when they think of a future where robots or computers will take their jobs, but it might be that there are areas where computers will never be better than us, such as creativity."[18]

The Penrose Institute has developed a number of chess puzzles that it claims humans can quickly unravel but which computers cannot, despite expending huge amounts of time and energy. It wants to study how it is that humans can come to the right conclusions so readily. James Tagg, who is leading the Penrose Institute, said: "We are interested in seeing how the eureka moments happen in people's brains. For me it is an actual flash of light but it will be different for others."[19]

God and humans

Where is God in all this? In much of the AI literature the answer is nowhere. That leaves me feeling uneasy. Not that I am a God-fearing member of society, let alone a signed-up adherent of any of the world's great religions. But if a group of technicians believes that they can venture into this most difficult territory of the relationship between mind and matter without reference to the contemplations of philosophers over the last 2,000 years and without even considering the religious view, I wonder if they can really appreciate the depth of the issues.

Penrose says that he is not a believer in God. But he has roundly criticized the arguments for atheism put forward by the late Sir Stephen Hawking and the distinguished biologist Richard Dawkins. And his view of the structure and nature of the universe is at least compatible with a theistic viewpoint.

Penrose says that he is not so much a dualist (a believer in the separate existence of mind and matter) as a triadist, believing that the nature of the universe (in the widest sense) is to be thought of as a sort of three-legged stool. The three legs are matter, mind (or consciousness), and eternal mathematical truths. He admits that we hardly understand anything about the interrelationships between these three and how mankind stands in relation to them. This is the subject of much of his recent work.

Among physicists, Penrose is a controversial figure. While deeply respecting his earlier work in physics and mathematics, many think that he is profoundly mistaken in his views about the physics of consciousness. Max Tegmark, whom we have encountered throughout this book and who is a professor of physics at MIT, is particularly critical.

As a mere economist, I am not in a position to debate the physics or mathematics underlying these disputes. Yet I do have a suspicion that Roger Penrose is on to something. Penrose himself admits that his ideas on the subject are still speculative. It seems to me that he could be wrong about many of the details of his speculations but right that consciousness is different and that we will only understand how it works and its relationship with the physical world once science has made another major leap forward.

If Penrose is substantially right, then I can see some of the consequences. For a start, the Singularity will never happen, and the dystopian vision of humanity's future described earlier in this chapter will never be realized. In that case, the vision of our economic future expounded in earlier chapters holds good without limit.

For Kevin Kelly and other AI enthusiasts, the implication of the further development of AI seems clearly to be a sense of human diminishment. I am not at all sure that this is right. Everything depends upon just what AI research achieves and what it doesn't, and upon what brilliant scientists like Roger Penrose discover about what consciousness really is, how it works, and how it interacts with the physical world.

If there comes to be widespread acceptance of something like Penrose's three-legged stool view, the result would be not a sense of human diminishment but rather a renewed sense of human self-confidence. There might even be an inching toward the belief that mind, in some form, is at the root of the universe and that we humans are deeply connected with the eternal.

Intriguingly, in a rather different way and with somewhat different implications, the AI visionary Ray Kurzweil comes to similar sounding conclusions. In 2015 he told his audience at Singularity University: "As we evolve, we become closer to God. Evolution is a spiritual process. There is beauty and love and creativity and intelligence in the world – it all comes from the neocortex. So, we're going to expand the brain's neocortex and become more godlike."[20]

Recognizing mind as a separate part of the universe, or even putting it in pole position, does not lead inevitably to belief in God. But after the materialism of the last few centuries it would be a major step in this direction. Wouldn't it be ironic if the search for superhuman artificial intelligence brought us face to face with the Almighty and Eternal?

Bibliography

Adams, D. (2009) *The Hitchhiker's Guide to the Galaxy*, London: Pan.

Aoun, J. E. (2017) *Robot-Proof: Higher Education in the Age of Artificial Intelligence*, Boston, MA: Massachusetts Institute of Technology.

Avent, R. (2016) *The Wealth of Humans: Work, Power, and Status in the Twenty-First Century*, London: Penguin Random House.

Baker, D. (2016) *Rigged: How Globalization and the Rules of the Modern Economy were Structured to Make the Rich Richer*, Washington, DC: Center for Economic and Policy Research.

Bootle, R. (2009) *The Trouble with Markets: Saving Capitalism from Itself*, London: Nicholas Brealey.

Bootle, R. (2017) *Making a Success of Brexit and Reforming the EU*, London: Nicholas Brealey.

Bostrom, N. (2014) *Superintelligence: Paths, Dangers, Strategies*, Oxford: Oxford University Press.

Bregman, R. (2017) *Utopia for Realists*, London: Bloomsbury.

Brockman, J. (2015) *What to Think about Machines That Think*, New York: HarperCollins.

Brynjolfsson, E. and McAfee, A. (2016) *The Second Machine Age: Work, Progress, and Prosperity in a Time of Brilliant Technologies*, New York: W. W. Norton & Company.

Caplan, B. (2018) *The Case Against Education: Why the Education System Is a Waste of Time and Money*, New Jersey: Princeton University Press.

Carr, N. (2010) *The Shallows*, New York: W. W. Norton & Company.

Chace, C. (2016) *The Economic Singularity*, London: Three Cs Publishing.

Cowen, T. (2013) *Average is Over*, New York: Dutton.

Darwin, C. (1868) *The Variations of Animals and Plants under Domestication*, London: John Murray.

Davies, P. (2019) *The Demon in the Machine*, London: Allen Lane.

Dawkins, R. (2006) *The God Delusion*, London: Penguin.

Diamond, J. (1997) *Guns, Germs and Steel*, London: Jonathan Cape.

Fisher, M. (1991) *The Millionaire's Book of Quotations*, London: Thorsons.

Ford, M. (2015) *The Rise of the Robots*, London: Oneworld.

Gordon, R. (2012) *Is US Economic Growth Over? Faltering Innovation Confronts Six Headwinds*, Cambridge, MA: National Bureau of Economic Research.

Gunkel, D. (2018) *Robot Rights*, Cambridge, MA: The MIT Press.

Harford, T. (2017) *Fifty Things that Made the Modern Economy*, London: Little Brown.

Harari, Y. N. (2011) *Sapiens: A Brief History of Humankind*, London: Harvill Secker.

Harari, Y. N. (2016) *Homo Deus: A Brief History of Tomorrow*, London: Harvill Secker.

Haskel, J. and Westlake, S. (2018) *Capitalism without Capital: The Rise of the Intangible Economy*, New Jersey: Princeton University Press.

Jaynes, J. (1990) *The Origin of Consciousness in the Breakdown of the Bicameral Mind*, New York: Houghton Mifflin.

Kelly, K. (2016) *The Inevitable: Understanding the 12 Technological Forces That Will Shape Our Future*, New York: Penguin.

Keynes, J. M. (1931) *Essays in Persuasion*, London: Macmillan.

Keynes, J. M. (1936) *General Theory of Employment, Interest and Money*, London: Macmillan.

Lawrence, M., Roberts C. and King, L. (2017) *Managing Automation*, London: IPPR.

Layard, R. (2005) *Happiness: Lessons from a New Science*, London: Allen Lane.

Leonhard, G. (2016) *Technology vs. Humanity: The Coming Clash between Man and Machine*, London: Fast Future Publishing.

Lin, P., et al. (2009) *Robots in War: Issues of Risks and Ethics*, Heidelberg: AKA Verlag.

Lowrey, A. (2018) *Give People Money*, New York: Crown.

Malthus, T. (1798) *An Essay on the Principle of Population*, London: J. Johnson.

Marx, K. and Engels, F. (1848) *Manifesto of the Communist Party*, London: Workers' Educational Association.

Maslow, A. (1968) *Toward a Psychology of Being*, New York: John Wiley & Sons.

Minsky, M. (1967) *Finite and Infinite Machines*, New Jersey: Prentice Hall.

Mokyr, J. (1990) *The Lever of Riches*, New York: Oxford University Press.

Morris, I. (2010) *Why the West Rules – for Now: The Patterns of History, and What They Reveal about the Future*, New York: Farrar, Straus and Giroux.

Pecchi, L. and Piga, G. (2008) *Revisiting Keynes: Economic Possibilities for Our Grandchildren*, Cambridge, MA: MIT Press.

Penrose, R. (1989) *The Emperor's New Mind*, Oxford: Oxford University Press.

Penrose, R. (1994) *Shadows of the Mind*, Oxford: Oxford University Press.

Piketty, T. (2014) *Capital in the Twenty-First Century*, Cambridge, MA: Harvard University Press.

Pinker, S. (2018) *Enlightenment Now: The Case for Reason, Science, Humanism, and Progress*, London: Allen Lane.

Pinker, S. (1994) *The Language Instinct*, London: Penguin.

Pistono, F. (2012) *Robots Will Steal Your Job But That's OK: How to Survive the Economic Collapse and Be Happy*, California: Createspace.

Polanyi, K. (1944) *A Short History of a "Family Security System,"* New York: Farrar & Rinehart.

Rawls, J. (1971) *A Theory of Justice*, Oxford: Oxford University Press.

Rifkin, J. (1995) *The End of Work*, New York: Putnam.

Roberts, C. and Lawrence, M. (2017) *Wealth in the Twenty-First Century*, London: IPPR.

Ross, A. (2016) *The Industries of the Future*, London: Simon & Schuster.

Say, J. (1803) *A Treatise on Political Economy*, New American Edition, 1859, Philadelphia: J.B. Lippincott & Co.

Schor, J. (1992) *The Overworked American: The Unexpected Decline of Leisure*, New York: Basic Books.

Schwab, K. (2018) *The Future of the Fourth Industrial Revolution*, London: Penguin Random House.

Scott, J. (2017) *Against the Grain: A Deep History of the Earliest States*, New Haven: Yale University Press.

Seldon, A. and Abidoye, O. (2018) *The Fourth Education Revolution*, Buckingham: University of Buckingham Press.

Shackleton, J. (2018) *Robocalypse Now?* London: Institute of Economic Affairs.

Shadbolt, N., and Hampson, R. (2018) *The Digital Ape*, London: Scribe.

Shanahan, M. (2015) *The Technological Singularity*, Cambridge: The MIT Press.

Simon, H. (1965) *The Shape of Automation for Men and Management*, New York: Harper.

Smith, A. (1776) *The Wealth of Nations*, London: William Strahan.

Stiglitz, J. E. (1969) *New Theoretical Perspectives on the Distribution of Income and Wealth among Individuals*, London: The Econometric Society.

Susskind, R. and Susskind, D. (2017) *The Future of the Professions: How Technology Will Transform the Work of Human Experts*, Oxford: Oxford University Press.

Tegmark, M. (2017) *Life 3.0: Being Human in the Age of Artificial Intelligence*, London: Allen Lane.

Templeton, J. (1993) *16 Rules for Investment Success*, San Mateo: Franklin Templeton Distributors, Inc.

Toffler, A. (1970) *Future Shock*, New York: Penguin Random House.

Van Parijs, P. and Vanderborght, Y. (2017) *Basic Income*, Cambridge: Harvard University Press Mass.

Voltaire (1759) *Candide*, Reprint 1991, New York: Dover Publications.

Wilde, O. (1888) *The Remarkable Rocket*, Reprint 2017 London: Sovereign Publishing.

Williams, T. (2003) *A History of Invention from Stone Axes to Silicon Chips*, London: Time Warner.

Wood, G. and Hughes, S., eds. (2015) *The Central Contradiction of Capitalism?*, London: Policy Exchange.

Notes

Preface

1 Reported in *The Daily Telegraph*, August 16, 2018.
2 Reported in the *Financial Times*, September 6, 2018.

Prologue

1 Gunkel, D. (2018) *Robot Rights*, Cambridge, MA: The MIT Press, p. ix.
2 Asimov, I. and Shulman, J.A. (1988) *Asimov's Book of Science and Nature Quotations*, New York: Grove Press.
3 Chace, C. (2016) *The Economic Singularity*, London: Three Cs Publishing, p. 208.
4 Bill Gates has said: "You cross the threshold of job-replacement of certain activities all sort of at once." "The result could be the eradication of whole classes of work at the same time – including warehouse work, driving, room clean-up." Quoted in the *Financial Times*, February 25/26, 2017. The late Sir Stephen Hawking said: "If machines produce everything we need, the outcome will depend on how things are distributed. Everyone can enjoy a life of luxurious leisure if the machine–produced wealth is shared, or most people can end up miserably poor if the machine-owners successfully lobby against wealth redistribution. So far, the trend seems

to be toward the second option, with technology driving ever-increasing inequality." Quoted by Barry Brownstein on CapX, March 21, 2018.

5 As reported by Rory Cellan-Jones, BBC technology correspondent, December 2, 2014.

6 He writes: "... by any definition of "thinking", the amount and intensity that's done by organic human-type brains will, in the far future, be utterly swamped by the cerebrations of AI. Moreover, the Earth's biosphere in which organic life has symbiotically evolved is not a constraint for advanced AI. Indeed, it is far from optimal – interplanetary and interstellar space will be the preferred arena where robotic fabricators will have the grandest scope for construction, and where non-biological 'brains' may develop insights as far beyond our imaginings as string theory is for a mouse." *The Daily Telegraph*, May 23, 2015.

7 Shanahan (2015).

8 Kurzweil is a striking and controversial figure, but he is far from alone. John Brockman has a similar vision. He has written: "If our future is to be long and prosperous, we need to develop artificial intelligence systems in the hope of transcending the planetary life cycles in some sort of hybrid form of biology and machine. So, to me, in the long term there's no question of 'us versus them.'" See Brockman, J. (2015) *What to Think About Machines That Think: Today's Leading Thinkers on the Age of Machine Intelligence* (New York: Harper Collins Publishers), p. 15.

9 Brockman, J. (2015) *What to Think about Machines That Think*, New York: HarperCollins, pp. 45–6.

10 Quoted by Brockman 2015, p. 362.

11 Ross, A. (2016) *The Industries of the Future*, London: Simon & Schuster, p. 35.

12 Anthes, G. (2017) Artificial Intelligence Poised to Ride a New Wave, *Communications of the ACM*, 60(7): p. 19.

13 See, for instance, the following: Owen-Hill, A. (2017) What's the Difference between Robotics and Artificial Intelligence?

https://blog.robotiq.com/whats-the-difference-between-robotics-and-artificial-intelligence, and Wilson H. (2015) What is a Robot Anyway?, *Harvard Business Review*, https://hbr.org/2015/04/what-is-a-robot-anyway, and Simon, M., (2017) What is a Robot?, https://www.wired.com/story/what-is-a-robot/, and Gabinsky, I. (2018) Autonomous vs. Automated, *Oracle Database Insider*. https://blogs.oracle.com/database/autonomous-vs-, and Cerf, V.G. (2013) What's a Robot?, *Association for Computing Machinery Communications of the ACM*, 56(1): p. 7.

Chapter 1

1 P. Krugman (2017) *New Zealand Parliament*, volume 644, week 63. https://www.parliament.nz/en/pb/hansard-debates/rhr/document/48HansD_20071204/volume-644-week-63-tuesday-4-december-2007.
2 Gordon, R. (2012) *Is US Economic Growth Over? Faltering Innovation Confronts Six Headwinds*, Working Paper: August, Massachusetts: NBER.
3 Actually, in his book published in 1817, David Ricardo warned that, for a time at least, the new technologies of the industrial Revolution could make workers worse off. According to Paul Krugman, modern scholarship suggests that this may indeed have happened for several decades.
4 Morris, I. (2010) *Why the West Rules – For Now: The Patterns of History, and What They Reveal About the Future*, New York: Farrar, Straus and Giroux, p. 492.
5 Critics will, I am sure, chastise me for showing the simple numbers in this chart, rather than using a log scale. In truth, I agonized over this choice, and also considered showing rolling 100-year increases. But the picture would not be that different, and in keeping with the book's objective of being readily accessible to the general reader, I was loath to

introduce such things as a log scale, which could easily put off many readers. So simplicity won out.

6 As you might expect, these figures are the subject of some controversy among economists. They come from the famous work of the economist Brad De Long, "Estimates of World GDP, One Million B.C.–Present", 1998, http://econ161.berkeley.edu/. They include the estimated benefits of new goods. (This is known in the literature as the "Nordhaus effect", after the economist William Nordhaus.) De Long also shows figures that exclude these benefits. The figure for the year 2000 is then only about 8½ times the figure for 1800.

7 There is a useful summary and discussion of the long historical trends in GDP and productivity in a Capital Economics study Vicky Redwood and Nikita Shah (2017) History Does Not Suggest Pessimism about Productivity Potential, *Capital Economics*, November, https://research.cdn-1.capitaleconomics.com/f993f5/history-does-not-support-pessimism-about-productivity-potential.pdf.

8 See Mokyr, J. (1990) *The Lever of Riches*, New York: Oxford University Press.

9 See Williams, T. (2003) *A History of Invention from Stone Axes to Silicon Chips*, London: Time Warner.

10 Scott, J. (2017) *Against the Grain: A Deep History of the Earliest States*, New Haven: Yale University Press.

11 This global picture looks slightly different for some key countries – but only slightly. The USA experienced some reasonable growth in per capita GDP in both the seventeenth and eighteenth centuries. The UK even managed growth in per capita GDP of as much as 0.3 percent per annum from the sixteenth century onward. But even in these two cases the growth rate paled into insignificance beside what was to come later.

12 Malthus, T. (1798) *An Essay on the Principle of Population*, London: J. Johnson.

13 Ibid.

14 Darwin, C. (1868) *The Variations of Animals and Plants under Domestication*, United Kingdom: John Murray.
15 See the article by Allen, R. C. (2009) Engels' Pause: Technical change, capital accumulation, and inequality in the British Industrial Revolution, *Explorations in Economic History*.
16 Harari, Y. N. (2016) *Homo Deus: A Brief History of Tomorrow*, London: Harvill Secker.
17 R. C. Allen, R. C. (2001) The Current Divergence in European Wages and Prices from the Middle Ages to the Frist World war, *Explorations in Economic History* 38, pp. 411–47.
18 Ricardo, D. (1821) *Principles of Political Economy and Taxation*.
19 Emily R. Kilby (2007) *The Demographics of the US Equine Population*, State of the Animals Series 4, Chapter 10, pp. 175–205. The Humane Society Institute for Science and Policy (Animal Studies Repository).
20 ONS (2013) *2011 Census Analysis, 170 Years of Industry*.
21 Figures taken from Ian Stewart, Debapratim De and Alex Cole (2014) *Technology and People: The Great Job-Creating Machine*, Deloitte.
22 See "Labour's Share," speech given by Andy Haldane, Chief Economist of the Bank of England, at the Trades Union Congress, London, on November 12, 2015. https://www.bankofengland.co.uk//media/boe/files/news/2015/november/labors-share-speech-by-andy-haldane
23 For historical data on economic growth, see Vicky Redwood and Nikita Shah (2017), op. cit.
24 I gave my analysis of the causes of the GFC in my book Bootle, R. (2009) *The Trouble with Markets*, London: Nicholas Brealey.
25 Admittedly, if you look at the recent growth figures for the world as a whole, things do not look too bad. In the years from 2008 to 2016, the growth of global GDP per capita averaged 2 percent per annum. This is slower than in the Golden Age from 1950 to 1973 and the period in the early 2000s when the emerging markets were roaring ahead, led by China. But it is faster than all other periods since 1500.

26 Yet this gives a totally misleading impression that all is well. The emerging markets have continued to grow pretty decently, albeit at much slower rates than before. When you look at the developed countries on their own, though, you see a very different picture. Indeed, in much of the developed world, productivity growth has fallen to almost zero. Since 2008 the growth of GDP per capita in the USA has been 0.6 percent, the lowest since the 1600s. In the UK it has been 0.4 percent, the lowest since the eighteenth century, and in Sweden it has been 0.7 percent, the lowest since the early nineteenth century. The British economists Nicholas Crafts and Terence Mills have estimated that in the early 1970s US Total Factor and Productivity (TFP), that is to say output per unit of capital and quality-adjusted labor (which is a measure of innovation or increased productivity that is not simply due to more factors (e.g., capital) being employed), was growing at just above 1.5 percent. It is now growing at about 0.9 percent.

27 See G. Grossman (2018) Growth, Trade and Inequality, *Econometrica*, 86(1): pp. 37–8.

28 As we saw in our brief review of ancient history above, it is not sufficient to generate rapid overall productivity growth for there to be rapid productivity growth in one sector of the economy. What also matters is the rate of productivity growth in the other sectors of the economy into which labor is released from the sector experiencing rapid productivity growth. So, it is possible that one sector, say IT, experiences rapid productivity growth and so do the sectors employing a good deal of IT, but if the labor released by the employment of IT systems finds employment in restaurants and care homes, where both the level and the growth rate of productivity is much lower, it is quite possible for the overall growth of productivity to fall back. This was the insight of the American economist William Baumol, presented in an article published in the 1960s. W. Baumol (1967) The Macroeconomics of Unbalanced Growth, *American Economic Review*, 57(3) (June): pp. 415–26.

29 But, of course, such a result is not bound to occur. It is more likely to occur the lower the sensitivity of demand for the output of the sector experiencing rapid productivity growth to lower prices (which will increase the amount of labor released into the rest of the economy), and the lower the rate of productivity growth in the rest of the economy. If the technological advance is big enough and its effects spread generally throughout the economy (as with electricity, for instance), then this negative result is unlikely to happen.

30 Accordingly, believing that it is happening now when it didn't in the first several decades of the twentieth century effectively amounts to the proposition that current and recent technological changes are simply not as significant as earlier ones. So the "Baumol" interpretation of the current technological slowdown really amounts to a subset of the school of economists who espouse "technological pessimism," that is to say, those who believe that, in the wider sweep of history, recent and current technological developments don't amount to very much.

31 Gordon, R. J. (2016) *The Rise and Fall of American Economic Growth*, USA: Princeton University Press.

32 Solow, R. (1987) We'd Better Watch Out *New York Times Book Review*, July 12, 1987.

33 Quoted in Brynjolfsson, E. and McAfee, A. (2016) *The Second Machine Age, Work, Progress, And Prosperity in a Time of Brilliant Technologies*, New York: W. W. Norton & Company, p. 112.

34 Feldstein, M. (2015) The US Underestimates Growth, USA: *Wall Street Journal*, May 18, 2015. Mind you, not all economists agree. A study by Byrne, D., Oliner, S., and Sichel, D. concludes the exact opposite. They reckon that the effect of correcting mismeasurement was to raise TFP growth in the tech sector and to reduce it everywhere else, with next to no net effect on the economy overall. See Bryne, D., Oliner, S., and Sichel, D., *Prices of High-Tech Products, Mismeasurements, and Pace of Innovation*, Cambridge, MA, National Bureau of Economic Research, 2017.

35 See Diamond, J. (1997*) Guns, Germs and Steel*, London: Jonathan Cape.
36 Romer, P. (2008) *Economic Growth* (Library of Economics and Liberty) http://www.econlib.org/library/Enc/Economicgrowth.html.

Chapter 2

1 Prime Minister of Canada at the World Economic Forum in Davos.
2 Rod Brooks gave four dollars per hour as the approximate cost of Baxter in response to a question at the Techonomy 2012 Conference in Tucson, Arizona, on November 12, 2012, during a panel discussion with Andrew McAfee.
3 Templeton, J. (1993) *16 Rules for Investment Success*, California: Franklin Templeton Distributors, Inc.
4 Rifkin, J. (1995) *The End of Work*, New York: Putnam Publishing Group.
5 Susskind, R. and Susskind, D. (2017) *The Future of the Professions: How Technology will Transform the Work of Human Experts*, Oxford: Oxford University Press, p. 175.
6 Quoted in Kelly, K. (2016, *The Inevitable: Understanding the 12 Technological Forces that Will Shape our Future*, New York: Penguin, p. 49.
7 This is known as Amara's law after the scientist Roy Amara. See Chace (2016) *The Economic Singularity*, London: Three Cs Publishing, pp. 76–7.
8 Chace, C. (2016) *The Economic Singularity*, London: Three Cs Publishing, p. 76.
9 Pistono, F. (2012) *Robots Will Steal Your Job But That's OK: How to Survive the Economic Collapse and Be Happy*, California: Createspace, p. 21.
10 For a discussion of machine learning, see Craig, C. (2017) "Machine Learning: The Power and Promise of Computers that Learn by Example," London: The Royal Society,

https://royalsociety.org/~/media/policy/projects/machine-learning-report.pdf

11 See Brockman, J. (2015) *What to Think about Machines that Think*, New York: Harper Collins Publishers, pp. 226–7.

12 *The Daily Telegraph*, December 23, 2015.

13 "Technological Growth and Unemployment: A Global Scenario Analysis," report of the *Journal of Evolution & Technology* (2014), https://jetpress.org/v24/campa2.htm

14 Aoun, J. E. (2017) *Robot-Proof: Higher Education in the Age of Artificial Intelligence*, USA: Massachusetts Institute of Technology, p. 1.

15 Tegmark, M. (2017) *Life 3.0 Being human in the age of Artificial Intelligence*, UK: Penguin Random House, p. 124.

16 Glenn, J. C., Florescu, E. and The Millennium Project Team (2016), http://107.22.164.43/millennium/2015-SOF-Executive Summary-English.pdf

17 Nedelkoska, L. and Quintini, G. (2018) *Automation, Skills Use and Training*, OECD Social, Employment and Migration. Working Papers 202, Paris: OECD Publishing, 2018, https://www.oecd-ilibrary.org/employment/automation-skills-use-and-training_2e2f4eea-en

18 Frey, C. B. and Osborne, M. A. (2013) "The Future of Employment: How Susceptible Are Jobs to Computerization?" https://www.oxfordmartin.ox.ac.uk/downloads/academic/The_Future_of_Employment.pdf

19 Quoted in the *Financial Times*, February 25/26, 2017.

20 Chui, M. Manyika, J. and Miremadi, M. (2015) "Four Fundamentals of Workplace Automation," *McKinsey Quarterly* (November).

21 Max Tegmark has laid down three criteria for judging whether a job is more or less likely to be challenged, or replaced, by robots any time soon. They amount to essentially the same as McKinsey's two criteria, with my suggested addition of "common sense." They are: Does it require interacting with people and using social intelligence? Does it involve creativity and coming up with clever solutions?

Does it require working in an unpredictable environment? Tegmark (2017), p. 121.

22 Chace, C. (2016).

23 Ibid., p. 249.

24 Simon, H. (1965) *The Shape of Automation for Men and Management*, New York: Harper.

25 Minsky, M. (1967) *Finite and Infinite Machines*, New Jersey: Prentice Hall.

26 Bostrom, N. (2014) *Superintelligence: Paths, Dangers, Strategies*, Oxford: Oxford University Press, p. 4.

27 According to Chace (2016), p. 14.

28 Reported in *The Economist*, April 21, 2018.

29 Markoff, J. (2012) How Many Computers to Identify a Cat? 16,000, *New York Times*, June 25, 2012.

30 Chace, C. (2016), p. 15.

31 Quoted in Autor, D. H. (2015) Why are there still so many jobs? The History and Future of Workplace Automation, *Journal of Economic Perspectives*, Vol. 29 (Summer 2015), p. 8.

32 Susskind and Susskind (2017), p. 276.

33 Ibid., pp. 272–3.

34 Shanahan, M. (2015) *The Technological Singularity*, Cambridge: The MIT Press, p. 162.

35 Quoted by Jeremy Warner in *The Daily Telegraph*.

36 Quoted in Kelly (2016), p. 176.

37 Haskel, J. and Westlake, S. (2017) *Capitalism Without Capital: The Rise of the Intangible Economy*, USA: Princeton University, p. 127.

38 See Autor, D. H. (2015).

39 Chace (2016), pp. 16–17.

40 Kelly (2016).

41 Avent, R. (2016) *The Wealth of Humans: Work, Power, and Status in the Twenty-First Century*, New York: St. Martin's Press, p. 59.

42 Ford, M. (2015) *The Rise of the Robots*, London: Oneworld, pp. 76–8.

Chapter 3

1 Simon, H. (1966) "Automation", letter in the *New York Review of Books*, May 26, 1966.
2 This remark, or something very much like it, is widely attributed to a range of people including Yogi Berra (the baseball coach of the New York Yankees), Niels Bohr, Albert Einstein, and Sam Goldwyn (the movie mogul).
3 This statement is widely attributed to the great sage but I have found it impossible to pin down chapter and verse.
4 Strictly speaking, there should be an increase in the volume of investment, but if the price of investment goods falls sufficiently then the total value of investment spending may not rise. This then undermines my subsequent point about real interest rates needing to rise.
5 Ford, M. (2015) *The Rise of the Robots*, London: Oneworld.
6 Say, J. B. (1803) *A Treatise on Political Economy*, New American Edition, 1859, Philadelphia: J.B. Lippincott & Co.
7 Downturns arise when, for a variety of reasons, the aggregate desire to save (i.e., not to spend all income) exceeds the aggregate desire to invest, and consequently aggregate demand falls short of productive potential. Demand bounces back when the aggregate desire to save falls short of the aggregate desire to invest.
8 Some economists argue that if we are again faced with a serious shortfall of demand, governments should do nothing. They should simply let demand fall short and let the economy turn down, and then recover through natural means. This is a return to what is called the classical view that was advocated by many economists in the 1930s. Those who take this line are sometimes described as the "Austrian school," after a group of Austrian economists, led by the late Friedrich von Hayek, who took this position. Some economists of this school argued that, if a depressed economy is left to its own devices, recovery will occur spontaneously as inefficient production is "purged." Others, typically from

the more analytical Anglo-Saxon tradition, also argued that recovery would occur automatically but because a depressed economy would cause prices to fall, thereby increasing the real value of the money supply, which would, ultimately, make people feel wealthier –this would spark an increase in spending and hence bring an economic recovery.

9 This is not the place to engage in a detailed discussion of Keynesian economics. Suffice it to say that many economists, myself included, think that the Austrian approach to depressions (although not necessarily to other things) is, to use a technical economic term, bonkers. What's more, the "neoclassical" approach that relies on an increasing real value of the money supply to bring recovery is bonkers squared.

10 Bootle, R. (2017) *Making a Success of Brexit and Reforming the EU*, London: Nicholas Brealey.

11 Reported in the *Financial Times*, September 6, 2018.

Chapter 4

1 Voltaire (1759) *Candide*. Reprinted in 1991, USA: Dover Publications.

2 This quotation is sometimes attributed to a number of different people, including Confucius.

3 Matthew 6:28, The Bible, New King James Version, Nashville: Thomas Nelson Inc., 1982.

4 Smith, A. (1776) *The Wealth of Nations*, London: William Strahan.

5 Marx, K. and Engels, F. (1848) *Manifesto of the Communist Party*, London: Workers' Educational Association.

6 Keynes, J. M. (1931) *Essays in Persuasion*, London: Macmillan.

7 There is an interesting collection of essays on Keynes edited by Pecchi, L. and Piga, G. (2008) *Grandchildren: Revisiting Keynes*, Cambridge, MA: MIT Press.

8 See Freeman, R. B. (2008) "Why Do We Work More than Keynes Expected?" in Pecchi and Piga, pp. 135–42.
9 J. E. Stiglitz (2010) "Toward a General Theory of Consumerism: Reflections on Keynes's Economic Possibilities for our Grandchildren," in L. Pecchi and G. Piga (2008), pp. 41–85.
10 Mokyr, J., Vickers, C. and Ziebarth, N. L. (2015) The History of Technological Anxiety and the Future of Economic Growth: Is this Time Different?, *Journal of Economic Perspectives*, Vol. 29 (Summer 2015). pp. 31–50.
11 Stiglitz (2010), op. cit.
12 Keynes (1931).
13 See Clark, A. and Oswald, A. J. (1994) Unhappiness and Unemployment, *Economic Journal*, Vol. 104, No. 424 (May), pp. 648–59.
14 Quoted by Freeman, R. B. (2008), op. cit.
15 Schor, J. (1992) *The Overworked American: The Unexpected Decline of Leisure*, New York: Basic Books), p. 47. It's worth noting that hunters and gatherers probably worked even less. Archeologists estimate their workweek at no more than 20 hours.
16 See Mokyr, Vickers, and Ziebarth (2015), op. cit.
17 Sources: England and Wales House Condition Survey (1967, 1976) and Rouetz, A. and Turkington, R. (1995), *The Place of the Home: English Domestic Environments, 1914–2000*, London: Taylor S. Francis.
18 Stiglitz (2010), op. cit.
19 "80 Percent Hate Their Jobs – But Should You Choose a Passion or Paycheck?" (2010) *Business Insider*, http://articles, businessinsider.com/2010-10-04/strategy30001895_1_new-job-passion-careers
20 Pistono, F. (2012) *Robots Will Steal Your Job But That's OK: How to Survive the Economic Collapse and Be Happy*, Scotts Valley: Createspace, pp. 135–6.
21 See Layard, R. (2005) *Happiness: Lessons from a New Science*, London: Allen Lane.
22 Williams, T. (2003) *A History of Invention from Stone Axes to Silicon Chips*, London: Time Warner.

23 Jerome, J. K. (1889) *Three Men in a Boat – To Say Nothing of the Dog!*, London: Penguin.
24 Reported in *The Daily Telegraph*, January 19, 2019.
25 In fact, we should not just lazily extrapolate from the very dramatic increases in life expectancy that have been achieved over the last hundred years. Over this period life expectancy has just about doubled. But, as Yuval Noah Harari has argued, this was not through the extension of a normal human life but rather because of a large reduction in the numbers of people suffering a premature death from malnutrition, infectious diseases and violence. In the distant past, if you escaped these depredations, it was not at all unusual for people to live to a ripe old age. As Harari (2016) points out, Galileo Galilei died at 77, Isaac Newton at 84, and Michelangelo at 88. He says: "In truth, so far modern medicine hasn't extended our natural life span by a single year."
26 Bregman, R. (2017) *Utopia for Realists,* London: Bloomsbury Publishing.
27 Wilde (1888), *The Remarkable Rocket*, Reprint 2017, London: Sovereign Publishing.
28 See Stiglitz (2010), op. cit.

Chapter 5

1 Gunkel, D. (2018) *Robot Rights*, Cambridge, Mass: The MIT Press, p. ix.
2 Quoted in Chace, C. (2016) *The Economic Singularity*, London: Three Cs Publishing.
3 Ross, A. (2016) *The Industries of the future*, London: Simon & Schuster, p. 130.
4 Ibid., p. 12.
5 Chace (2016), op. cit., pp. 117–18.
6 The World Economic Forum (2018) *Reshaping Urban Mobility with Autonomous Vehicles*, Geneva: World Economic Forum.

7 Quoted in R. Dingess (2017) Effective Road Markings are Key to an Automated Future, *Top Marks (The Magazine of Road Safety Markings and Association)*, Edition 19.
8 Speaking at the Royal Society in London, reported in *The Daily Telegraph*, May 14, 2018.
9 Schoettle, B. and Sivak, M. I. (2015) *A Preliminary Analysis of Real-World Crashes Involving Self-Driving Vehicles*, The University of Michigan Transportation Research Institute, Report No. UMTRI-2015-34, October.
10 Reported in *The Daily Telegraph*, May 5, 2018.
11 *Financial Times*, December 3, 2018, p. 20.
12 Dingess, R. (2017), op. cit.
13 See BikeBiz website http://bit.ly/2maBbno.
14 For a skeptical view of the prospects for driverless vehicles, see Christian Wolmar, "False Start," *The Spectator*, July 7, 2018 and his book (2017) *Driverless Cars: On a Road to Nowhere*, London: London Publishing Partnership.
15 Wikipedia (2018) "Military Robot," https://en.wikipedia.org/wiki/Military_robot.
16 P. Lin et al. (2009) *Robots in War: Issues of Risks and Ethics*, AKA Verlag Heidelberg, pp. 51–2.
17 Unmanned Effects (UFX), *Taking the Human Out of the Loop*, U.S. Joint Forces Command Rapid Assessment Process Report, prepared by Project Alpha, 2003, p. 6.
18 Singer, P. (2000) Robots at War: The New Battlefield, *The Wilson Quarterly*, adapted from *Wired for War: The Robotics Revolution and Conflict in the twenty-first Century*, London: Penguin Press, 2009, available at https://wilsonquarterly.com/quarterly/winter-2009-robots-at-war/robots-at-war-the-new-battlefield/
19 This information comes from Cowen (2013).
20 Pinker, S. (1994) *The Language Instinct*, London: Penguin, pp. 190–1.
21 Reported in *The Daily Telegraph*, December 31, 2018.
22 Reported in *The Daily Telegraph*, January 22, 2018.

23 Harford, T. (2017) *Fifty Things that made the Modern Economy*, London: Little Brown.
24 Reported in the *Financial Times*, June 25, 2018.
25 Chace (2016), op. cit., pp. 252–3.
26 Ford, M. (2015) *The Rise of the Robots: Technology and the Threat of Mass Unemployment*, Great Britain: Oneworld publications, pp. 123–4.
27 World Economic Forum, in collaboration with The Boston Consulting Group (2018) *Toward a Reskilling Revolution A Future of Jobs for All*, Geneva: World Economic Forum.
28 Ford (2015), op. cit., p. 162.
29 See the report in *The Daily Telegraph*, February 26, 2018.
30 Referred to in Ross (2016), op. cit., p. 33.
31 Quoted in Chace (2016), op. cit., p. 146.
32 Referred to in Susskind, R. and Susskind, D. (2017) *The Future of the Professions: How Technology will Transform the Work of Human Experts*, Oxford: Oxford University Press, pp. 45–7.
33 Adams 2009.
34 Chace (2016), op. cit., p. 165.
35 Keynes, J. M. (1936) *The General Theory of Employment, Interest and Money*, London: Macmillan.

Chapter 6

1 Harari, Y. N. (2011) *Sapiens: A Brief History of Humankind*, London: Harvill Secker.
2 Quoted on Icahn, C.'s Twitter feed, https://twitter.com/carl_c_icahn?lang=en
3 Case, A. and Deaton, A (2015) Rising Morbidity and Mortality in Midlife among White Non-Hispanic Americas in the Twenty-first Century, *PNAS*, 112(49), Princeton: Woodrow Wilson School of Public and International Affairs and Department of Economics, Princeton University, Princeton, September 17.

4 Bregman, R. (2017) *Utopia for Realists*, London: Bloomsbury Publishing, p. 185.
5 Income inequality between quintiles has not been as high; it did not rise as much as in the USA during the 1980s, and it has been fairly unchanged since 1990. Between 1980 and 2014 real disposable income in the UK rose by 86 percent. Post-tax incomes of the top quintile doubled, while those of the bottom quintile increased by 62 percent. The top income quintile earned 37 percent of total post-tax income in 1980, with 5 percent going to the top 1 percent. By 1990 these shares had risen to 43 percent and 8 percent, respectively. Since 1990, the top quintile share has not changed very much, while the top 1 percent share continued to rise until 2007. These data come from World Inequality Database, https://wid.world/data/ and Office for National Statistics, Effects of taxes and benefits on household income: historical datasets, https://www.ons.gov.uk/peoplepopulationandcommunity/personalandhouseholdfinances/incomeandwealth/datasets/theeffectsoftaxesandbenefitsonhouseholdincomehistoricaldatasets
6 Quoted by Schwab, K. (2018) *Shaping the Future of the Fourth Industrial Revolution*, Penguin Radom House: London, p. 23.
7 Kelly, K. (2012) Better than Human: Why Robots Will – and Must – Take Our Jobs, *Wired*, December 24, 2012, p. 155.
8 Brynjolfsson, E. and McAfee, A. (2016) *The Second Machine Age, Work, Progress, And Prosperity in a Time of Brilliant Technologies*, New York: W. W. Norton & Company, p. 157.
9 Ibid., p. 179.
10 Piketty, T. (2014) *Capital in the Twenty-First Century*, Massachusetts: Harvard University Press.
11 See "Thomas Piketty's *Capital*, Summarised in Four Paragraphs," *The Economist*, May 2014, Lawrence Summers, "The Inequality Puzzle, *Democracy: A Journal of Ideas*, No. 33 (Summer 2014; Mervyn King, "*Capital in the Twenty-First Century* by Thomas Piketty," review, *The Daily Telegraph*, May 10, 2014.

12 See M. Feldstein in G. Wood and Steve Hughes, (eds) (2015) *The Central Contradiction of Capitalism?*, London: Policy Exchange.
13 Grubel, H. in Wood and Hughes (2015), op. cit.
14 Giles, C. in Wood and Hughes (2015), op. cit.
15 Stiglitz, J. E. (1969) *New Theoretical Perspectives on the Distribution of Income and Wealth among Individuals*, London: The Econometric Society.
16 See Sargent, J. R. in Wood and Hughes (2015) op. cit.
17 Haskel, J. and Westlake, S. (2018) *Capitalism without Capital: The rise of the intangible economy*, USA: Princeton University Press, pp. 127–8.
18 Kelly, K. (2012), op. cit.
19 Avent, R. (2016) *The Wealth of Humans*, UK: Penguin Random House, p. 51.
20 Autor, D. (2015) Why are There Still so Many Jobs? The History and Future of Workplace Automation, *The Journal of Economic Perspectives*, 29(3), pp. 3–30.
21 Reported in the *Financial Times*, January 29, 2018.
22 Lawrence, M. Roberts, C. and King, L. (2017) *Managing Automation*, London: IPPR.
23 Chace (2016), pp. 51–2.
24 A report from the Center for American Entrepreneurship, referred to in an article by John Thornhill in the *Financial Times*, October 23, 2018.
25 International Federation of Robotics (IFR).
26 Goldman Sachs (2017) "China's Rise Artificial Intelligence," August 31, 2017.
27 House of Lords (2018) Committee Report, session 2017-19, HL Paper 100, April 16, p. 117.
28 Ibid., p. 117.
29 Chace, C. (2016).
30 IMF(2018), Manufacturing Jobs: Implications for Productivity and Inequality, chapter 3 of *World Economic Outlook*, Washington, DC: IMF, April.

Chapter 7

1 L. Floridi (2017) Robots, Jobs, Taxes, and Responsibilities, *Philosophy & Technology*, March, 30(1), pp. 1–4.
2 Maslow, A. (1968) *Toward a Psychology of Being*, New York: John Wiley & Sons.
3 Kelly, K. (2016) *The Inevitable: Understanding the 12 Technological Forces that Will Shape our Future*, New York: Penguin, p. 190.
4 Reported in the *Financial Times*, September 6, 2018.
5 Delaney, K. J. (2017) The robot that takes your job should pay taxes, says Bill Gates, *Quartz*, February 17, https://qz.com/911968/bill-gates-the-robot-that-takes-your-job-should-pay-taxes.
6 Walker, J. (2017) Robot Tax – A Summary of Arguments 'For' and 'Against,' *Techemergence*, October 24, 2017, https://www.techemergence.com/robot-tax-summary-arguments/.
7 Isaac, A. and Wallace, T. (2017) Return of the Luddites: why a robot tax could never work, *The Daily Telegraph*, September 27, 2017, https://www.telegraph.co.uk/business/2017/09/27/return-luddites-robot-tax-could-never-work/.
8 Walker, J. (2017), op. cit.
9 Ibid.
10 Reuters (2017) European Parliament calls for robot law, rejects robot tax, February 16, https://www.reuters.com/article/us-europe-robots-lawmaking/european-parliament-calls-for-robot-law-rejects-robot-tax-idUSKBN15V2KM
11 See Abbott, R. and Bogenschneider, B. (2017) Should Robots Pay Taxes? Tax Policy in the Age of Automation, *Harvard Law & Policy Review*, Vol. 12, p. 150.
12 L. Summers (2017) Robots are wealth creators and taxing them is illogical, *Financial Times*, March 5.
13 R. J. Shiller (2017) "Robotization without Taxation?," *Project Syndicate*, http://prosyn.org/Rebz6Jw.
14 Tegmark, M. (2017) *Life 3.0: Being Human in the Age of Artificial Intelligence*, London: Allen Lane., p. 273.

15 Dvorsky, G. (2017) Hackers have already started to weaponize artificial intelligence, *Gizmodo*, November 9, https://gizmodo.com/hackers-have-already-started-to-weaponize-artificial-in-1797688425.

16 Ibid.

17 House of Lords Committee Report, session 2017-19, HL Paper 100, April 16, 2018, p. 95.

18 Quoted in (2016) *Robotics and Artificial Intelligence*, House of Commons Science and Technology Committee, Fifth Report of Session 2016-17, HC 145, London: House of Commons.

19 Ibid., p. 18.

20 For an analysis of these issues, see (2017) *Data Management and Use: Governance in the Twenty-first Century*, London: British Academy and the Royal Society, https://royalsociety.org/~/media/policy/projects/data-governance/data-management-governance.pdf.

21 Globe editorial (2018) When tech companies collect data, bad things can happen, January 30, https://www.theglobeandmail.com/opinion/editorials/globe-editorial-when-tech-companies-collect-data-bad-things-can-happen/article37798038/

22 Baker, P. (2018) Reining In Data-Crazed Tech Companies, April 16, https://www.ecommercetimes.com/story/85278.html

23 Solon, O. (2018) Facebook says Cambridge Analytica may have gained 37m more users data, *The Guardian*, April 4, https://www.theguardian.com/technology/2018/apr/04/facebook-cambridge-analytica-user-data-latest-more-than-thought

24 Frischmann, B. (2018) Here's why tech companies abuse our data: because we let them, *The Guardian*, April 10, https://www.theguardian.com/commentisfree/2018/apr/10/tech-companies-data-online-transactions-friction

25 Stucke, M. E. (2018) Here are all the reasons it's a bad idea to let a few teach companies monopolize our data, *Harvard Business Review*, March 27, https://hbr.org/2018/03/

here-are-all-the-reasons-its-a-bad-idea-to-let-a-few-tech-companies-monopolize-our-data

26 Mintel (2018) Data Danger: 71 Percent of Brits Avoid Creating New Company Accounts Because of Data Worries, 23 May http://www.mintel.com/press-centre/technology-press-centre/data-danger-71-of-brits-avoid-creating-new-company-accounts-because-of-data-worries

27 EU GDPR (2018) https://www.eugdpr.org/eugdpr.org-1.htm

28 Johnston, I. (2018) EU funding 'Orwellian' artificial intelligence plan to monitor public for 'abnormal behaviour,' *The Daily Telegraph*, September 19, https://www.telegraph.co.uk/news/uknews/6210255/EU-funding-Orwellian-artificial-intelligence-plan-to-monitor-public-for-abnormal-behavior.html

29 Arun, C. (2017) AI Threats to Civil Liberties and Democracy, speech in Berkman Klein Centre for Internet & Society, October 1, transcript available at http://opentranscripts.org/transcript/ai-threats-civil-liberties-democracy/

30 Lucas, L. and Feng, E. (2018) Inside China's Surveillance State," *Financial Times*, July 20, https://www.ft.com/content/2182eebe-8a17-11e8-bf9e-8771d5404543

31 Vincent, J. (2018) Artificial intelligence is going to supercharge surveillance, *The Verge*, January 23, https://www.theverge.com/2018/1/23/16907238/artificial-intelligence-surveillance-cameras-security

32 Zeng, M. J. (2018) China's Social Credit System puts its people under pressure to be model citizens, The Conversation, January 23, https://theconversation.com/chinas-social-credit-system-puts-its-people-under-pressure-to-be-model-citizens-89963

33 Ibid.

34 Lucas and Feng (2018) Inside China's surveillance state, *The Financial Times* July 20, https://www.ft.com/content/2182eebe-8a17-11e8-bf9e-8771d5404543, accessed on August 14, 2018.

35 *The South China Morning Post* reported in *The Daily Telegraph*, May 1, 2018.
36 Shadbolt, N. and Hampson, R. (2018) *The Digital Ape*, London: Scribe.
37 P. Domingos (2015) *The Master Algorithm: How the Quest for the Ultimate Learning Machine Will Remake Our World*, New York: Basic Books.
38 Polonski, V. (2017) "How artificial intelligence silently took over democracy," World Economic Forum, August 12, https://www.weforum.org/agenda/2017/08/artificial-intelligence-can-save-democracy-unless-it-destroys-it-first/
39 J. M. Burkhardt (2017) How Fake News Spreads, *Library Technology Reports*, 53(8).
40 Polonski, V. (2017) Artificial intelligence has the power to destroy or save democracy, Council on Foreign Relations, August 7, https://www.cfr.org/blog/artificial-intelligence-has-power-destroy-or-save-democracy
41 BBC News, Fake news a democratic crisis for UK, MPs warn, July 28, 2018, https://www.bbc.co.uk/news/technology-44967650
42 Reported in *The Guardian*, January 22, 2018.
43 *Wired* (2018), https://www.wired.com/story/emmanuel-macron-talks-to-wired-about-frances-ai-strategy/

Chapter 8

1 Joi, J. I. (2016) Society-in-the-Loop," Massachusetts: MIT Media Lab, August 12.
2 Quoted in Leonhard, G. (2016) *Technology vs. Humanity: The Coming Clash between Man and Machine*, London: Fast Future Publishing, p. 60.
3 Goel, A. K. (2017) AI Education for the World, *AI Magazine*,38(2), pp. 3–4.
4 Wohl, B. (2017) Coding the curriculum: new computer science GCSE fails to make the grade, *The Conversation*, June 21,

http://theconversation.com/coding-the-curriculum-new-computer-science-gcse-fails-to-make-the-grade-79780

5 V. Matthews (2018) Teaching AI in schools could equip students for the future, *Raconteur*, May 23, https://www.raconteur.net/technology/ai-in-schools-students-future

6 Kosbie, D. et al. (2017), op. cit.

7 TEALS, https://www.tealsk12.org

8 Matthews, V. (2018), op. cit.

9 Williamson, B. (2017) "Coding for What? Lessons from Computing in the Curriculum," talk prepared for the Pop Up Digital Conference, Gothenburg, Sweden, June 19.

10 G. Brown-Martin, (2017) Education and the Fourth Industrial Revolution, literature review prepared for Groupe Média TFO, August, p. 4, https://www.groupemediatfo.org/wp-content/uploads/2017/12/FINAL-Education-and-the-Fourth-Industrial-Revolution-1-1-1.pdf

11 Quoted in Leonhard 2016, p. 24.

12 See Kosbie, D. Moore, A. W. and Stehlik, M. (2017) How to Prepare the Next Generation for Jobs in the AI Economy, *Harvard Business Review*, June 5 https://hbr.org/2017/06/how-to-prepare-the-next-generation-for-jobs-in-the-ai-economy

13 Brown-Martin, G. (2017) op. cit.

14 Aoun, J. E. (2017) *Robot-Proof: Higher Education in the Age of Artificial Intelligence*, USA: Massachusetts, Institute of Technology, p. xviii.

15 Ibid., p. 51.

16 Seldon, A. and Abidoye, O. (2018) *The Fourth Education Revolution*, Buckingham: University of Buckingham Press.

17 Ford, M. (2015) *The Rise of the Robots*, London: Oneworld, p. 146.

18 See Susskind, R. and Susskind, D. (2017) *The Future of the Professions: How Technology will Transform the Work of Human Experts*, Oxford: Oxford University Press.

19 *The Guardian*, March 22, 2016.

20 K. Robinson, *Ed Tech Now*, January 20, 2012.

21 Quoted in Seldon (2018).

22 Michael Gove speech at the BETT Show 2012, available at https://www.gov.uk/government/speeches/michael-gove-speech-at-the-bett-show-2012
23 Bootle, R. (2012) *The Trouble with Markets: Saving Capitalism from Itself*, London: Nicholas Brealey.
24 Caplan, B. (2018) *The Case Against Education*, Princeton: Princeton University Press.
25 Ibid.
26 Quoted by Foroohar, R. in the *Financial Times*, November 12, 2018.
27 Quoted in Seldon and Abidoye (2018).
28 For a discussion of this, and other educational issues, see Chao Jr., R. (2017) Educating for the Fourth Industrial Revolution, *University World News*, No. 482 November 10, http://www.universityworldnews.com/article.php?story=20171107123728676 and Brown-Martin, G. 2017, op. cit.
29 Toffler, A. (1970) *Future Shock*, New York: Penguin Random House.
30 Carr, N. (2010) *The Shallows*, New York: W. W. Norton & Company.
31 Brockman, J. (2015), pp. 26–7.
32 Autor, D. (2015) Why Are There Still So Many Jobs? The History and Future of Workplace Automation, *The Journal of Economic Perspectives*, 29(3).

Chapter 9

1 Leonhard, G. (2016), p. 49.
2 This parody of synthetic logic has wide applicability. Its origin is unknown.
3 The most recent attempt to produce a grand vision of a fair society has come from the American philosopher John Rawls, in his book (1971) *A Theory of Justice*, Oxford: Oxford University Press.
4 Quoted in Bregman (2017) *Utopia for Realists*, p. 72.

5 See Baker, D. (2016) *Rigged: How Globalisation and the Rules of the Modern Economy were Structured to Make the Rich Richer, Washington D.C.:* Center for Economic and Policy Research.
6 For a wide-ranging survey of all aspects of this subject, see L. Martinelli, "Assessing the Case for a Universal Basic Income in the UK," IPR Policy Brief, September 2017, University of Bath, and also OECD, *Basic Income as a Policy Option: Can It Add Up?*, Paris: OECD.
7 See *Our Common Wealth: a Citizens' Wealth Fund for the UK*, London: IPPR, 2018, http://www.ippr.org/research/publications/our-common-wealth
8 *Financial Times*, July 2, 2018.
9 Quoted in Lowrey, A. (2018) *Give People Money*, New York: Crown).
10 Quoted in Van Parijs and Vanderborght, Y. (2017) *Basic Income*, Cambridge: Harvard University Press Mass, p. 79.
11 Polanyi, K. (1944) *A Short History of a Family Security System*, New York: Farrar & Rinehart.
12 Bregman (2017), pp. 88–9.
13 Ibid., p. 97.
14 J. K. Galbraith, "The Unfinished Business of the Century," Lecture given at the London School of Economics, June 1999.
15 See Van Parijs, P. and Vanderborght, Y. (2017) *Basic Income*, Cambridge: Harvard University Press Mass.
16 Turner, A. "Capitalism in the Age of Robots: Work, Income, and Wealth in the Twenty-First Century," lecture given at the School of Advanced Studies, Johns Hopkins University, Washington, DC, April 20, 2018.
17 Bootle, R. (2009) *The Trouble with Markets: Saving Capitalism from Itself*, London: Nicholas Brealey.
18 Chace (2016), pp. 217–18.
19 Lowrey, A. (2018) *Give People Money: The simple idea to solve inequality and revolutionise our lives*, London: WH Allen.
20 Referred to in Lowrey, A. (2018).
21 Quoted in Van Parijs and Vanderborght (2017), p. 85.
22 Ibid.
23 "The Basics of Basic Income," www.johnkay.com

24 Piketty, T. (2013) *Capital in the Twenty-First Century*, USA: Harvard University Press.
25 Pinker, S. (2018) *Enlightenment Now*, London: Allen Lane.

Conclusion

1 This remark is widely credited to the American baseball coach and philosopher Yogi Berra, but it or something like it is also attributed to a wide range of other people.
2 Sagan, C. (1980) *Cosmos*, New York: Random House.

Epilogue

1 See Darrach, B., Meet Shaky, the First Electronic Person, *Life*, November 20, 1970, p. 68.
2 Quoted by Brockman (2015), p. 166.
3 Ford (2015), pp. 229–30.
4 Leonhard (2016), p. 9.
5 This would have a consequence of key interest not so much to economists as to our cousins, the accountants. (Distant cousins, several times removed, I must stress.) If AI develops to the point where machines can continually improve themselves, this will reduce, or perhaps even reverse, the process of deterioration that affects all physical assets, which gives rise to the accounting concept of "depreciation," which plays such a large part in many companies' accounts. At least in some cases, could "depreciation" turn into "appreciation"?
6 Kelly (2016), p. 49.
7 Tegmark (2017), p. 314.
8 Brockman (2015), pp. 29–30.
9 Take this, for example, from Shanahan, M.: "Foremost among these limitations is mortality. An animal's body is a fragile thing, vulnerable to disease, damage, and decay, and the biological brain, on which human consciousness (today)

depends, is merely one of its parts. But if we acquire the means to repair any level of damage to it, and ultimately to rebuild it from scratch, possibly in a non-biological substrate, then there is nothing to preclude the unlimited extension of consciousness" (2015, p. xxi).

10 Ford (2015), pp. 230–2.

11 Quoted in Shanahan (2015), p. 157. Shanahan expands on this vision. He has written:
"Unhampered by earthly biological needs, capable of withstanding extremes of temperature and doses of radiation that would be fatal to humans, and psychologically untroubled by the prospect of thousands of years traveling through interstellar space, self-reproducing super intelligent machines would be in a good position to colonize the galaxy. From a large enough perspective, it might be seen as human destiny to facilitate this future, even though (unenhanced) humans themselves are physically and intellectually too feeble to participate in it."

12 As Nick Bostrum puts it: "The existence of birds demonstrated that heavier-than-air flight was physically possible and prompted efforts to build flying machines. Yet the first functioning airplanes did not flap their wings. The jury is out on whether machine intelligence will be like flight, which humans achieved through an artificial mechanism, or like combustion, which we initially mastered by copying naturally occurring fires" (2014, p. 34).

13 Harari (2016), pp. 140–1.

14 Jaynes, J. (1990) *The Origin of Consciousness in the Breakdown of the Bicameral Mind*, New York: Houghton Mifflin.

15 Dawkins, R. (2006) *The God Delusion*, London: Penguin.

16 Shanahan, M. (2015) *The Technological Singularity*, Cambridge: The MIT Press, p. 98.

17 Brockman (2015), p. 255.

18 Quoted in *The Daily Telegraph*, March 14, 2017. For a detailed exposition of Penrose's views on these issues, see Penrose 1989 and 1994.

19 Quoted in *The Daily Telegraph*, March 14, 2017.

20 Quoted in Leonhard (2016), p. 9.

Index

3D printing 37

Africa 190
aggregate demand 23, 29, 32–3, 73–80
agriculture 18, 19, 25–6
AI (artificial intelligence) 1–4, 5–9, 39, 276–8, 280–4
  aggregate demand 75–7
  comparative advantage 64–5
  cost 61–3
  crime 207–8
  deterrence 196–7
  development 43–7, 66–7
  economics 10–11, 73, 74
  education 220, 221, 222–4, 225, 229–32, 236–9, 242–3
  employment 123–4, 158–9, 171–3, 178–80
  encouragement 197–8
  ethics 291–2
  exponential growth 52–3
  healthcare 149–52
  income distribution 253–4
  inflation 80–1, 82
  legal framework 209–11, 212
  liberty 212–14
  limitations 57–9
  location 183–7
  macroeconomics 70–2
  negativity 42
  politics 214–16
  productivity 84–5
  research restrictions 205–7, 217
  Singularity 285–9
  taxation 199–204
  terrorism 208–9
  underperformance 53–5
  unemployment 48–51
aircraft 136, 137–8, 142
Al-Khalili, Jim 85, 197–8
Alaska 268
aliens 4–5
Allen, Robert 23
AlphaGo 46
Amazon 144
Andreessen, Marc 47
Ansip, Andrus 199
Aoun, Joseph 224
Aristotle 48
Arun, Chinmayi 213
Ashton, Kevin 46–7
Asimov, Isaac 122, 204–5
assets 90–2, 93
Autor, David 176, 241

Baker, Pam 211
banking 79, 147, 175
Barra, Hugo 148
Bart Everett, H. R. 143
Baxters 62
Bean, Sir Charles 35–6

benefits *see* welfare state
Bentham, Jeremy 248, 264, 291
Bernanke, Ben 75
Berners-Lee, Sir Tim 215
biotechnology 37, 79
birth rate 22
BJG (basic jobs guarantee) 258–9
Blade Runner (film) 55
Blair, Tony 232–3
Bletchley Park 43
bonds 91
books 60–1
bots 216
Breedlove, Philip M. 143
Bregman, Rutger 122, 163, 264
Brin, Sergey 132
Brockman, John 4, 6, 287, 290
Brown-Martin, Graham 224
Brynjolfsson, Erik 166
Buffett, Warren 276
business 8, 11

Callaghan, James 59
Cambridge Analytica 211–12, 215
Canada 267
capital 20, 21, 71–2, 88–9, 177–8
  investment 200–1
  Piketty 169
capitalism 101, 260, 261, 262
Caplan, Bryan
  The Case against Education 234, 235, 240
car industry 26, 55, 73; *see also* driverless vehicles
Carr, Nicholas
  The Shallows 239
censorship 148
Centre for Cities, The 182
Chace, Calum 1, 45, 52, 182, 266
Chakrabarti, Shami 213
chemical weapons 206
chess 46, 57, 63, 292–3
children 8, 11–12, 21, 120
China 31, 61, 78, 81–2
  AI investment 185, 191
  censorship 148
  labor costs 189, 190
  mass surveillance 213–14
Chomsky, Noam 57, 288
Christianity 99, 293–4
Clark, Donald 230
coal industry 162
common sense 51
communism 101, 247
comparative advantage 64–5
competition 106–8
computerization 31, 34, 43, 47–8, 60–1, 183
  education 221–3
  see also internet
consciousness 289–90, 291–3
construction industry 20, 144
Coplin, Dave 210
Corbyn, Jeremy 198
creativity 51, 52
Creemers, Rogier 213–14
Cuban, Mark 223
culture 187–8
cybercrime 207–9

da Vinci robots 150–1
Darwin, Charles 22
data 211–12
Dawkins, Richard 289, 293
debt 80
Deep Blue 46, 57
DeepMind 185–6
defense 188–9
DeLong, Brad 18
demand deficiency 77–8

democracy 214–15
Dershowitz, Alan 166
development ladder 190–1
Dewey, John 229
Diamond, Jared 36
digital economy 35–6, 82–3, 165
Dingess, Robert 136
domestic services 145–7
dot-com boom 41, 66
driverless vehicles 128–36, 137, 138–42, 174, 209
drones 137–8, 143
drudgery 100, 101
Dvorsky, George P. 208

e-readers 60–1
economics 9, 10–11, 18–20, 29–31
  downturns 32–6, 74–6
  forecasting 69–70
  growth 93
  Industrial Revolution 15–17, 22–5, 27–8
  unemployment 72–3
  wages 63–4
  see also GDP; macroeconomics
Edinburgh 180–1
education 9, 11–12, 120–1, 219–24, 233–6
  AI 230–2, 236–9, 242–3
  leisure 226–7
  methods 228–30
  reform 224–6
  social mobility 254
  the state 239–42
  tertiary 232–3
Einstein, Albert 59
elections 211–12, 215–16
electricity 37
emotional intelligence 51, 52
employment 8, 11, 19, 97–9, 161–3
  AI 123–4, 144–5, 147–52, 158–9, 178–80
  attitudes 99–102
  competition 106–8
  domestic 146–7
  education 235–6, 241
  enjoyment 108–9
  forecasting 127–8
  hours 104–5, 110–11, 113–15, 117–20, 125–6
  human skill 51–2, 171–3, 175–6
  income 124–5
  labor costs 189–91
  leisure 153–4
  life balance 102–3
  public policy 196–7
  purpose 116–17
  status 115–16
  technology 25–7
  UBI 263
  wages 63–4
  see also unemployment
Engels, Friedrich 23
equities 91
eternal life 2, 3
ethics 204–5, 206, 291–2
eugenics 206, 289
European Union (EU) 198–9, 205
exponential growth 44–5, 52–3

Facebook 211–12, 216
fake news 215–16
Fan Hui 46
Feldstein, Martin 36, 168
ferries 141
Fields, Mark 132
finance 31–2, 36, 180–2; *see also* banking
Finland 267–8
fintech industry 175

First Agricultural Revolution 19
Florida, Richard 184
food 22
Ford, Henry 26
Ford, Martin 64–5, 228
France 198, 268
Franklin, Benjamin 101
Frey, Carl 50
Friedman, Milton 75, 261, 270–1
Friedman, Tom 183

Galbraith, John Kenneth 69, 261, 265
gambling 156
gaming 46
Gates, Bill 2, 171, 198, 276
GDP (gross domestic product) 16–17, 18, 20–2, 30, 38
  growth 85–6
  UBI 265, 266, 273
  under-recording 35–6
general-purpose technologies (GPTs) 44
Ghosn, Carlos 136
Giles, Chris 168
Global Financial Crisis (GFC) 31–2, 36, 61, 67
globalization 61, 81–2, 162, 163–4, 183–4
GMI *see* UBI
Go 46
Goldin, Ian 182
Gordon, Robert 33–5, 39, 42
Gove, Michael 230
government 9, 11, 75, 196–8, 210
  cybercrime 208–9
  education 239–42
  income distribution 249, 254–5
  see also public policy; taxation; welfare state
Great Depression 23, 29, 67
Great Recession 31–2, 36, 61, 67
Grubel, Herbert 168

Hammond, Philip 129
Hamon, Benôit 198
happiness 109, 113, 248
Harari, Yuval Noah 23, 289
Harford, Tim 145
Hassabis, Demis 133
Hawking, Sir Stephen 2, 3, 4, 293
Hayek, Friedrich von 261
healthcare 149–52, 282; *see also* surgery
Hillman, Nick 232
holidays 119–20, 125
Hong Kong 189
horses 25–6
human race 3, 5, 6, 48, 154–7, 280–2
  consciousness 289–91, 292–3
  liberty 212–14
  Singularity 286–8

immigration 256–7
income 12, 27, 74
  distribution 77–8, 79–80, 124–5, 171–2, 246–50, 253–6
  inequality 76–7
  taxation 199–200, 202
  technology 173–6
  see also UBI
INDECT 213
India 189, 190
Industrial Revolution 15–17, 18, 22–5, 27–8, 33–4, 38, 66
  steam engine 36–7
  work 99, 100–1
inequality 107, 163–71, 255–6, 275–6

AI 173–6, 178–80
education 240–1
welfare state 254–5
inflation 30, 80–3
inheritance 249–50
insurance 209–10, 252
intelligence 289, 290
interest rates 80, 87–91, 93
internet 26, 31, 43, 46–7, 66, 215
investment 29, 33, 41, 71–2, 155
robots 62, 184–5
Iran 268
Ishikawa, Masatoshi 187–8
Italy 269

Japan 188, 189
Jaynes, Julian 289, 290
Jevons, William Stanley 127
Johnston, Ian 213

Kasparov, Gary 46, 57
Kelly, Kevin 46, 62, 171, 197, 287, 294
Keynes, John Maynard 74–5, 125
General Theory 155–6
"The Economic Possibilities for Our Grandchildren" 102–4, 105, 107, 109, 111, 117
Kosbie, David 224
Kurzweil, Ray 3, 294
The Singularity is Near: When Humans Transcend Biology 285, 288
Kuznets, Simon 179

Lane, Dale 221
Lawrence, D. H. 100–1
legal framework 209–11
legal profession 104, 108, 114, 147, 175
leisure 26, 34, 48, 110–11, 121–3
education 226–7
employment 153–4
holidays 119–20, 125
increased 86–7, 93–4, 98, 282–3
location 181
work balance 113–15
Leonhard, Gerd 223–4
liberty 212–14
living standards 16, 19, 21–3, 28, 37–8, 42
increase 86–7, 103–4
Keynes 102–3
London 180–2
Lowrey, Annie 268
Luckin, Rose 222
Luddites 25

McAfee, Andrew 166
McCarthy, John 6
McDonnell, John 259, 269
McKinsey 49–51
macroeconomics 11, 67, 70–2, 92–4
Macron, Emmanuel 216
Malthus, Thomas 21–2, 264
malware 207
manufacturing 26, 162, 191
Markoff, John 53
Marshall, Alfred 165
Marshall Plan 29
Marx, Karl 99, 101, 247, 264
mass surveillance 212–14
material needs 111–12, 124–5
May, Theresa 152
mechanization 25, 47
medical science 37, 151–2
military applications 142–3
Mill, John Stuart 261
"Millennium Project, The" 49
Minsky, Marvin 54

Mokyr, Joel 50
money 74–5
Moore, Andrew 197
Moore's Law 45, 52
Moravec, Hans 288
Moravec, Peter 56
More, Sir Thomas 260
Musk, Elon 131, 140, 260

nanotechnology 37, 79
Native Americans 268
Nazis 43, 206, 289
Neumann, John von 285
Newcomen, Thomas 36
Nixon, Richard 261
nominal interest rates 89–90
nuclear weapons 138, 206

Obama, Barack 215
Ocado 145
OECD 50, 51–2
oil 29, 30, 78
Osborne, Michael 50

Paine, Thomas 260
parenting 8
Pearson, Allison 54
Pearson, Ian 45
Penrose, Sir Roger 292, 293–4
Petrov, Stansilav 138
physical labor 100
Piketty, Thomas:
  Capital in the Twenty-First Century 167–71, 275
Pinker, Steven 144, 275, 288
Pistono, Federico 45
Polanyi, Karl 56
  The Great Transformation 263–4
Poole, David 201
population 20–2, 77–8
positional goods 107
poverty 12, 275
power 108, 115
privacy 132, 139, 211, 212–13
productivity 19, 26–8, 32–4, 35–6, 38, 42
  AI 44
  capital 71–2
  growth 83–6
property 91–2, 147
public policy 5, 9, 10, 79–80, 88, 196–7
  robots 204–5
  UBI 272–4
purpose 116–17

railways 140–1
RAND Corporation 101
Rees, Martin, Lord 3, 4, 5
regionality 180–6
relationships 157
religion 99, 187–8, 293–4
retail 112, 144–5
retirement 8, 121, 122
Reuther, Walter 73
Ricardo, David 29, 64, 189, 264
  Principles of Political Economy and Taxation 25
Rifkin, Jeremy
  The End of Work 41–2
Robinson, Sir Ken 225, 229
robots 1–2, 4, 8–9, 39, 280–4
  aggregate demand 75–7
  comparative advantage 64–5
  cost 61–3
  culture 187–8
  definition 6, 7
  deterrence 196–7
  development 66–7
  economics 10–11, 73, 74

employment 178–80
ethics 291–2
growth 45
healthcare 149–51
inflation 80–1
investment 184–5
macroeconomics 70–2
military applications 142–3
negativity 42
productivity 44, 48, 84–5
public policy 204–5
taxation 198–204, 216
underperformance 55–6
Rodrik, Dani 190
Romer, Paul 37
Ross, Alec 129
Rowling, J. K. 165
R.U.R. (Čapek) 6
Russell, Bertrand 260
Russell, Stuart 4

Samuelson, Paul 261
savings 88–9
Say's Law 74
Scarpetta, Stefano 51–2
Schor, Juliet 111
Schumpeter, Joseph 24
Scott, James 19
Searle, John 58
Second World War 29–30, 142
secular stagnation 32
Seldon, Sir Anthony 228, 230
service industry 26, 44, 84–5, 191; *see also* education; healthcare
sex 157
Shanahan, Murray 3, 58, 290
Shaw, George Bernard 100, 101
Shiller, Robert 202
shortage 82, 83
Simon, Herbert 54, 59
Simpson, O. J. 166
Singapore 189
Singularity, The 3–4, 5–6, 285–9, 294
skilled labor 172–3, 175–6
slavery 48, 115, 163, 209, 260
Smith, Adam 29
The Wealth of Nations 28, 99
Solow, Robert 34
South Korea 189, 198, 206
state, the *see* government; welfare state
status 115–16, 117
steam engine 36–7, 44
Stiglitz, Joseph 32, 169
Stoll, Cliff 43
Summers, Larry 32, 259
surgery 66, 84, 150–1, 166
Switzerland 181, 236–7, 269

Tagg, James 293
Taiwan 189
Takahashi, Professor 58
taxation 9, 168, 252, 253
robots 198–204
UBI 257, 270–1, 272
teachers 221–2, 228–9, 231–2
TEALS program 221–2
technology 19–20, 22, 24–5, 26–8, 38
advances 37
cognitive ability 239
emerging markets 31
GPT 44
income 173–6
inequality 164–5
monopoly 176–8
negativity 41–2, 59–61
postwar 29
slowdown 33–4

technology (*cont.*)
underestimation 42–3
see also AI; computerization; robots
Tegmark, Max 49, 52, 287, 293
Templeton, Sir John 41
terrorism 139, 208–9
"Thatcher Revolution" 162
Thiel, Peter 34
Tobin, James 261
Tocqueville, Alexis de 264
Toffler, Alvin 238
trade 24, 28, 29, 78, 189–90
trains 140–1
translation services 148
Tufekci, Zeynep 211
Turing, Alan 43, 287
Turner, Adair, Lord 265–6

Uber 129, 173–4
UBI (universal basic income) 203, 204, 254, 256–62, 269–74, 277, 283
opposition 263–4, 266–7
side benefits 264–6
tests 267–9
UBS (universal basic services) 258, 259
unemployment 1, 72–3, 109, 162, 270
AI 48–51
government spending 203–4
Industrial Revolution 23, 24–5
inflation 30, 82–3
technology 41–2
see also welfare state
United Kingdom (UK) 185–6, 198, 269; *see also* London
universities 232–3, 235, 236–7
unskilled labor 172–3, 175–6
Urmson, Chris 131

Van Parijs, Philippe 271
Vanderborght, Yannick 271
Varian, Hal 171
Vinge, Vernor 285

war 29; *see also* military applications; Second World War
Watson, Thomas J. 43
Watson (machine) 46, 58
Watt, James 36, 37, 44
wealth 12, 116, 247–8, 255, 276
Piketty 167, 168, 169–70
weaponry 206
welfare state 9, 250–3, 255, 261–2, 263–4
UBI 267, 273
Wilde, Oscar 122
Williamson, Ben 222
Winfield, Alan 210
winner-takes-all markets 165–6
women 105
work *see* employment

Zuckerberg, Mark 260